Springer Tracts in Mechanical Engineering

About this Series

Springer Tracts in Mechanical Engineering (STME) publishes the latest developments in Mechanical Engineering - quickly, informally and with high quality. The intent is to cover all the main branches of mechanical engineering, both theoretical and applied, including:

- Engineering Design
- Machinery and Machine Elements
- Mechanical structures and Stress Analysis
- Automotive Engineering
- Engine Technology
- Aerospace Technology and Astronautics
- Nanotechnology and Microengineering
- Control, Robotics, Mechatronics
- MEMS
- Theoretical and Applied Mechanics
- Dynamical Systems, Control
- Fluids mechanics
- Engineering Thermodynamics, Heat and Mass Transfer
- Manufacturing
- Precision engineering, Instrumentation, Measurement
- Materials Engineering
- Tribology and surface technology

Within the scopes of the series are monographs, professional books or graduate textbooks, edited volumes as well as outstanding PhD theses and books purposely devoted to support education in mechanical engineering at graduate and post-graduate levels.

Indexed by SCOPUS and Springerlink.

To submit a proposal or request further information, please contact: Dr. Leontina Di Cecco Leontina.dicecco@springer.com or Li Shen Li.shen@springer.com.

Please check our Lecture Notes in Mechanical Engineering at http://www.springer.com/series/11236 if you are interested in conference proceedings. To submit a proposal, please contact Leontina.dicecco@springer.com and Li.shen@springer.com.

More information about this series at http://www.springer.com/series/11693

Saad Kashem • Romesh Nagarajah •
Mehran Ektesabi

Vehicle Suspension Systems and Electromagnetic Dampers

Springer

Saad Kashem
Swinburne University of
Technology Sarawak
Kuching, Sarawak
Malaysia

Romesh Nagarajah
Swinburne University of Technology
Melbourne, VIC
Australia

Mehran Ektesabi
Swinburne University of Technology
Melbourne, VIC
Australia

ISSN 2195-9862 ISSN 2195-9870 (electronic)
Springer Tracts in Mechanical Engineering
ISBN 978-981-10-5477-8 ISBN 978-981-10-5478-5 (eBook)
DOI 10.1007/978-981-10-5478-5

Library of Congress Control Number: 2017946769

Printed on acid-free paper

This Springer imprint is published by Springer Nature
The registered company is Springer Nature Singapore Pte Ltd.
The registered company address is: 152 Beach Road, #21-01/04 Gateway East, Singapore 189721, Singapore

Author's Declaration

I hereby declare that I am the sole author of this manuscript. To the best of my knowledge, the document contains no material previously published or written by another person except where due reference is made in the text.

Dr. Saad Kashem

Acknowledgements

It is a pleasure to thank all the people who made this possible. It is my great pleasure to offer warm thanks to Professor Saman Halgamuge who is the assistant dean of the Melbourne School of Engineering at the University of Melbourne. The effort and time he took to help me to validate the designed full car analytical model were outstanding.

I have been privileged to work with and learn from Timothy Barry and Mehedi Al Emran Hasan. They helped me to learn MATLAB/Simulink. I am also grateful to them for helping me to get through the difficult times and for all the emotional support.

It is my pleasure to thank Jason Austin, Simon Lehman and Alex Barry who worked with me to set up and experiment the Quanser suspension plant. From my supervision of their undergraduate final-year project on active suspension system, I have learned many things.

I wish to thank Dr. Durul Huda for his time and patience in teaching me about the dynamics of the full car model. I would like to thank the many people who have taught me science, including my high school teachers (especially Abdul High) and my undergraduate faculties at East West University (especially Md. Ishfaqur Raza PhD, Dr. Ruhul Amin, Dr. Anisul Haque, Dr. Mohammad Ghulam Rahman, Dr. Khairul Alam, Dr. Tanvir Hasan Morshed), for their wise advice, helping with various applications and so on.

Lastly, and most importantly, I wish to thank my parents. They supported me and loved me. To them I dedicate this book.

And special thanks to almighty Allah who made this book possible.

Abstract

A suspension system is an essential element of a vehicle to isolate the frame of the vehicle from road disturbances. It is required to maintain continuous contact between a vehicle's tyres and the road. In order to achieve the desired ride comfort and road handling performance, many types of research have been conducted. A new modified skyhook control strategy with an adaptive gain that dictates the vehicle's semi-active suspension system is presented. The proposed closed-loop feedback system first captures the road profile input over a certain period. Then it calculates the best possible value of the skyhook gain for the subsequent process. Meanwhile, the system is controlled according to the new modified skyhook control law using an initial or previous value of the skyhook gain. In this book, the proposed suspension system is compared with passive and three other recently reported skyhook controlled semi-active suspension systems through a virtual environment with MATLAB/Simulink as well as an experimental analysis with Quanser suspension plant. Its performances have been evaluated in terms of ride comfort and road handling performance. The model has been validated in accordance with the international standards of admissible acceleration levels ISO2631 and human vibration perception. This control strategy has also been employed on the full car model to improve the isolation of the vibration and handling performance of the road vehicle.

This book also describes the development of a new analytical full vehicle model with nine degrees of freedom, which uses the new modified skyhook strategy to control the full vehicle vibration problem. Nowadays, many researchers are working on active tilting technology to improve vehicle cornering. But in those work, the effect of road bank angle is not considered in the control system design or in the dynamic model of the tilting standard passenger vehicles. The non-incorporation of road bank angle creates a non-zero steady-state torque requirement. Therefore, in this manuscript, this phenomenon was addressed while designing the direct tilt control and the dynamic model of the full car model.

This book has indicated the potential of the SKDT suspension system in improving cornering performances of the vehicle and paves the way for future work on vehicle's integrated system for chassis control.

Keywords Quarter-car • Vehicle • Suspension • Semi-active • Skyhook • Adaptive • Control • Damper • Quanser

Contents

List of Figures

List of Tables

Chapter 1
Introduction

Abstract In this chapter, background of this book has been described. Motivation and methodologies has been depicted in the later section. A brief outline of this manuscript has been included in the last section.

1.1 Background

One of the most important considerations of the present automotive industry is to provide passenger safety, through optimal ride comfort and road holding, for a large variety of vehicle manoeuvres and road conditions. The comfort and safety of the passenger travelling in a vehicle can be improved by minimizing the body vibration, roll and heave of the vehicle body through an optimal road contact for the tyres. The system in the vehicle that provides these actions is the vehicle suspension, i.e. a complex system consisting of various arms, springs and dampers that separate the vehicle body from the tyres and axles (Fig. 1.1). In general, vehicles are equipped with fully passive suspension systems due to their low cost and simple construction. The passive suspension consists of springs, dampers and anti-roll bars with fixed characteristics. The major drawback of the passive suspension design is that you cannot simultaneously maximize both vehicle ride and handle performance. To achieve better ride performance, a "soft" suspension needs to be introduced to maintain contact between the vehicle body and the tyre. The "soft" suspension easily absorbs road disturbances. That is why most of the luxury cars employ "soft" suspensions to provide a comfortable ride. The second characteristic of vehicle performance is the road handling. This refers to a vehicle's ability to maintain contact between the vehicle's tyre and the road during turns and other dynamic manoeuvres. This can be achieved by "stiff" suspensions as seen in sports cars. The challenge of the passive suspension system is in achieving the right compromise between the two characteristics of vehicle performance which will best suit the targeted consumer. However, by introducing the active or semi-active suspension system in the vehicle (Fig. 1.2), a more desirable compromise can be achieved between the benefits of the soft and stiff suspension system.

The active or semi-active suspension systems are incorporated with the active components, such as actuators and semi-active dampers, coupled with various

S. Kashem et al., *Vehicle Suspension Systems and Electromagnetic Dampers*,
Springer Tracts in Mechanical Engineering, DOI 10.1007/978-981-10-5478-5_1

Fig. 1.1 Rear suspension system without wheel of a vehicle

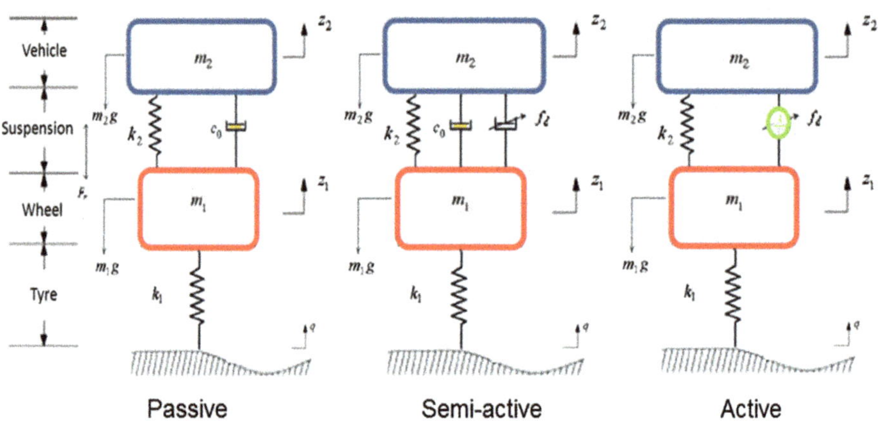

Fig. 1.2 The passive, semi-active and active suspension system

dynamic control strategies. With active components, these systems can provide adjustable spring stiffness and damping coefficients adapted to various road conditions.

Since the early 1970s, many types of active and semi-active suspension systems have been proposed to achieve better control of damping characteristics. Although the active suspension system shows better performance in a wide frequency range, its implementation complexity and cost prevent wider commercial applications. That is why the semi-active suspension system has been widely studied to achieve high levels of performance in terms of vehicle suspension system. To control the damper of the semi-active suspension system, many control strategies including skyhook surface sliding mode control [1], neural network control [2], H-infinity control [3], skyhook control, ground hook control, hybrid control [4, 5], fuzzy logic control [6, 7], neural network-based fuzzy control [8], neuro-fuzzy control [9], discrete time fuzzy sliding mode control [10], optimal fuzzy control [11], and adaptive fuzzy logic control [12, 13] have been explored. Between all of the above control systems, the skyhook control proposed by Karnopp et al. in 1974 [14] is widely used since it yields the best compromise between vehicle performance and practical implementation of semi-active suspension systems.

In the past few decades, researchers have modified the basic skyhook control strategy by adding some variations and have named them optimal, modified or adaptive type skyhook control strategies [15, 16]. But in most of these studies, skyhook gain (SG) of the control strategy remains as a constant value, and it is usually chosen from a set of values as suited for the vehicle in the simulation environment. One of the major goals of this manuscript is to present a new modified skyhook control strategy with adaptive SG.

This control strategy has also been employed on the full car model to improve the isolation of the vibration and handling the performance of the road vehicle. The full car model designed in this manuscript has nine degrees of freedom, and those are the heave modes of four wheels and the heave, lateral, roll, pitch and yaw modes of the vehicle body.

Nowadays, some researchers have focused on active steering control to improve vehicle cornering [17–19]. Three types of active steering control strategies have been proposed. These are the four-wheel active steering system (4WAS), the front wheel active steering system (FWAS) and the active rear wheel steering system (RWAS). The four-wheel active steering system (4WAS) is the combination of the rear active steering system and the front active steering system. In the FWAS system, the front wheel steer angle is determined by the steering angle generated due to the driver's direct steering input and a resultant corrective steering angle input that is produced by the design of the active front wheel steering controller.

Vehicle performance during cornering has been improved by most of the car manufacturers by using electronic stability control (ESC). Car manufacturers use different brand names for ESC, such as Volvo named it DSTC (Dynamic Stability and Traction Control); Mercedes and Holden called it ESP (Electronic Stability Program); and DSC (Dynamic Stability Control) is the term used by BMW and Jaguar, but despite the term used, the processes are almost the same. To avoid oversteering and understeering during cornering, ESC extends the brake and different torque on each wheel of the vehicle. But ESC reduces the longevity of the tyre as the tyre skids while random braking. To overcome this problem, a vehicle can be tilted inward via an active or semi-active suspension system.

The concept of "active tilting technology" has become quite popular in narrow tilting road vehicles and modern railway vehicles. Now in Europe, most new high-speed trains are fitted with active tilt control systems, and these trains are used as regional express trains [20, 21]. To tilt the train inward during cornering, tilting actuators are used as an element of the secondary active suspension system. These actuators are named as bolsters. In a road vehicle, actuators are also used to affect the vehicle roll angle via an active suspension system. Since the beginning of the 1950s, there has been extensive work done in developing the narrow tilting vehicle by both the automotive industry [22–25] and academic researchers [26–30].

This particular small and narrow geometric property of the vehicle poses stability problems when the vehicle needs to corner or change a lane. There are also two types of control schemes that have been used to stabilize the narrow tilting vehicle. These control schemes are defined as direct tilt control (DTC) and steering tilt control (STC) systems as detailed in [27, 31, 32]. A typical passenger vehicle body can be tilted up to ten degrees as the maximum suspension travel is around 0.25 m. Then, the lateral acceleration of the tilted vehicle caused by gravity can reach a maximum of about 0.17 g [33]. Since the lateral acceleration produced by normal steering manoeuvres is around 0.3–0.5 g, the active or semi-active suspension systems have the potential of improving vehicle ride handling performance [33]. Semi-active or active suspension systems can act promptly to tilt the vehicle with the help of semi-active dampers or actuators. However, the active suspension systems need to avoid over-sensitive reaction to driver's steering commands for vehicle safety. Recently Bose Corporation presented the Bose suspension system [34] in which the high-bandwidth linear electromagnetic dampers improved vehicle cornering. It is able to counter the body roll of the vehicle by stiffening the suspension while cornering. Car giant Nissan has developed a four-wheeled ground vehicle named Land Glider [35]. The vehicle body can lean into a corner up to 17° for sharper handling considering the speed, steering angle and yaw rate of the vehicle. In addition, in the works stated above and other research, the effect of road bank angle is neither considered in the control system design nor in the dynamic model of the tilting standard passenger vehicles [26, 27, 31, 32, 36–44]. Not incorporating the road bank angle creates a non-zero steady-state torque requirement. So this phenomenon needs to be addressed while designing the tilt control and the dynamic model of the full car model. To lean a vehicle which incorporates the road bank angle, the response time of the actuator or semi-active damper plays an important role.

The majority of the semi-active suspension systems use pneumatic or hydraulic solutions as the actuator or semi-active damper [45–49]. These systems are characterized by high force and power densities but suffer from low efficiencies and response bandwidths. Commercial systems incorporating electromagnetic elements (combine rotary actuators and mechanical elements) illustrate the properties of the magneto-rheological fluids in damper technology to provide adjustable spring stiffness. However, linear electromagnetic actuators appear as a better solution for a semi-active suspension system in respect of their high force densities, form factor, and response bandwidth. The motivation and the methodology of this manuscript are described in the next section.

1.2 Motivation and Methodologies

The active suspension system has exploited superior performance in terms of vehicle ride comfort and ride handling performances compared to other passive and semi-active suspension systems in the automotive industry. Nevertheless, they are not widely commercialized yet because of their high cost, weight, complexity and energy consumption. Another major drawback of the active suspension system is that it is not fail-safe in the situation of a power breakdown. That is why the semi-active suspension system has been widely studied and commercialized to achieve high levels of performance with ride comfort and road handling. To control the damper of the semi-active suspension system, many control strategies have been proposed, but among all of them, skyhook control proposed by Karnopp et al. in 1974 [14] is widely used since it yields the best compromise between vehicle performance and practical implementation of semi-active suspension systems. The skyhook control system has been adopted and implemented to offer superior ride quality to commercial passenger vehicles. However, this technology is still an emerging one, and elaboration and more research work on different theoretical and practical aspects are required. In the past few decades, researchers have modified the basic skyhook control strategy by adding some variations and naming them optimal, modified or adaptive type skyhook control strategy [15, 16]. But in most of these studies, skyhook gain (SG) of the control strategy remains as a constant value, and it is usually chosen from a set of values as suited for the vehicle in the simulation environment. One of the major goals of this book is to present a new modified skyhook semi-active control strategy with adaptive skyhook gain.

According to this strategy, each wheel of the car behaves independently. At first, the road profile input has been captured for each wheel from the tyre deflection measurements over a certain period of time. Then the quarter-car model is simulated onboard computer of the vehicle. It follows the new modified skyhook control strategy with a range of SG. This method determines a certain value of SG which is applied to the new modified skyhook control strategy to dictate the semi-active suspension system of the corresponding car wheel. Meanwhile, the system behaves according to the modified skyhook control law with an initial or previous value of the SG. After each period of time, SG is updated to match the road disturbance.

To evaluate the performance of the proposed closed-loop feedback system, two degrees of freedom quarter-car model has been used. The vibration isolation and road handling performance of the proposed model have been analysed and compared with a passive system and three other skyhook controlled systems subject to base excitation defined by ISO ISO8608 [50]. The other control systems are the continuous skyhook control of Karnopp et al. [14], the modified skyhook control of Bessinger et al. [15] and the optimal skyhook control of Nguyen et al. [16]. An experimental evaluation of the proposed skyhook control strategy has also been done by the Quanser quarter-car suspension plant. Then the control strategy has been employed on the full car model to improve the isolation of the vibration and handling the performance of the road vehicle. The full vehicle model designed in this manuscript has nine degrees of freedom: the heave modes of four wheels and the heave, lateral, roll, pitch and yaw modes of the vehicle body.

Another major objective of this manuscript is to improve the performance of vehicles during cornering with little or no skidding using a new approach. That approach tilts the standard passenger vehicle inward during cornering or sudden lane change with consideration of the road bank angle, the steering angle, lateral position acceleration, yaw rate and the velocity of the vehicle. The suspension system considered here consists of the linear electromagnetic damper (LEMD) in parallel with the conventional mechanical spring and damper. This manuscript has two goals, firstly to find out the possibilities of tilting a car inward through a semi-active suspension system and secondly to improve the vehicle ride comfort and road handling performance. The stability control algorithm for tilting vehicles has been designed in such a way that the driver does not need to have special driving skills to operate the vehicle. In this manuscript, the shortcomings of existing direct tilt control systems are addressed. At first, a dynamic model of a tilting vehicle which considers the road bank angle is designed. Then an improved direct tilt control system along with the modified skyhook control system design is presented. This system takes into account the steering angle, the road bank angle, lateral position acceleration, yaw rate and the velocity of the vehicle. A yaw-rate sensor and a lateral acceleration sensor are placed at the vehicle. The job of these sensors is to monitor the movement of the car body along the vertical axis. The combined control system will do a comparative analysis of the target value calculated and the actual value based on the driver's input through the steering. Then control system will make a decision considering the road bank angle, lateral position acceleration, yaw rate and velocity of the vehicle. The moment the car begins to turn, the control system will intervene by applying a precisely metered electromagnetic force using the separate linear electromagnetic damper placed at each wheel. This lifts up the side of the vehicle's body opposite to the centre of the turn and turns down the side which is on the same side of the turning point. This will make a certain angle between the vehicle body and the road as directed by the controller. This angle, between the road and the vehicle body, will move the vehicle's centre of gravity towards the turning point and will help the driver to turn smoothly using less road surface. Moreover, it will support the vehicle as it turns with more speed without skidding. This manuscript does not develop a new semi-active suspension physical model or a linear electromagnetic damper. The application of semi-active suspension with linear electromagnetic suspension system is suggested due to their reliability and effectiveness over other technology and for practical implementation.

To achieve the manuscript objectives, this research makes effective use of different analysis methods, including MATLAB/Simulink simulation processes and real-time tests and experiments where applicable. The next section outlines the structure of the whole manuscript.

1.3 Outline

Following this introduction chapter, the remainder of the manuscript is divided into seven more chapters. Chapter 2 includes an extensive review of the literature on different types of semi-active suspension control systems. Five widely known control approaches are reviewed more deeply. Since the damper plays an important role in the semi-active suspension system design, different types of damper technologies are discussed including Quanser electromagnetic damper which has been used in the experimental analysis of this manuscript. Also described is the tilting vehicle technology designed and developed by both the automotive industry and academic researchers.

In Chap. 3, the vehicle suspension system is categorized and discussed briefly. High- and low-bandwidth suspension system is also discussed. This chapter also examines the uncertainties in modelling a quarter-car suspension system caused by the effect of different sets of suspension parameters of a corresponding mathematical model. From this investigation, a set of parameters were chosen which showed a better performance than others in respect of peak amplitude and settling time. These chosen parameters were then used to investigate the performance of a new modified continuous skyhook control strategy as set out in Chap. 4.

Chapter 4 consists of a brief discussion on the proposed modified skyhook control approach, optimal skyhook control of Nguyen et al. [51], modified skyhook control of Bessinger et al. [15] and continuous skyhook control of Karnopp et al. [14]. A road profile was generated to study the performance of the different controllers. The two degrees of freedom quarter-car model described in Chap. 3 was simulated to compare the controller's performances. Quanser quarter-car suspension plant has been also used to compare the performance of the controllers in the experimental environment. These models have also been evaluated in terms of human vibration perception and admissible acceleration levels based on ISO 2631 in this chapter.

Chapter 5 presents a methodology on how to integrate the proposed skyhook control in a full car model to improve ride comfort and handling via a semi-active suspension system. A technique to determine the vehicle rollover propensity to avoid tipping over is also described. The road profile and four driving scenarios are discussed in this chapter briefly which form a basis for the analysis described in the next two chapters. A method to determine the admissible acceleration level based on ISO 2631 is also discussed in this chapter. The next chapter contains the simulation results of the semi-active suspension system developed as described in this chapter.

In Chap. 6, the analysis of the simulation results of the dynamic model of a full car model which considers the road bank angle is presented. The first section describes the parameters of the full car that were used in the analysis model and the environment of the simulation. The second section describes the performance of the proposed skyhook control system under different road conditions. In the third section, the performance of the combined approach, the proposed skyhook controller activated with the direct tilt control, is evaluated in different driving scenarios. The next section is comprised of the summary of the simulation while the vehicle is travelling on road class C and following driving scenario four.

In Chap. 7, the analysis of the dynamics of a full car model is presented. It incorporates the response of the Quanser quarter-car suspension plant as one of the four wheels of the full car model. The performance of the combined approach where the proposed skyhook controller is activated along with the direct tilt control is evaluated in Sects. 7.2 and 7.3 at frequency domain and time domain.

Chapter 8 presents the overall conclusion of this book, followed by future research recommendations.

Chapter 2
Control Strategies in the Design of Automotive Suspension Systems

Abstract In the literature available, many robust and optimal control approaches or algorithms were found in the design of automotive suspension systems. In this chapter, some of these will be reviewed such as the linear time-invariant H-infinity control (LTIH), the linear parameter-varying control (LPV) and model-predictive controls (MPC). Five widely known control approaches, namely, the linear quadratic regulator (LQR) and linear quadratic Gaussian (LQG), sliding mode control (SMC), fuzzy and neuro-fuzzy control, skyhook and groundhook approaches, are reviewed more deeply. Since the damper plays an important role in the semi-active suspension system design, different types of damper technologies are discussed in the second section. This includes the Quanser electromagnetic damper that was used in the experimental analysis in this manuscript. Another major objective of this manuscript is to tilt the standard passenger vehicle inward during cornering. So a brief literature review on automotive tilting technology is included in the last section.

2.1 Control Strategies

In general, a controlled system consists of a plant with sensors and actuators, and a control method is called a semi-active control strategy. A semi-active system is a compromise between the active and passive systems. It offers some essential advantages over the active suspension systems. The active control system depends entirely on an external power source to control the actuators and supply the control forces. In many active suspension applications, this control approach needs a large power source. On the other hand, semi-active devices need a lot less energy than the active ones. Another critical issue of the active control system is the stability robustness problem with respect to sensors or the whole system failure; this issue becomes a big concern when centralized controllers are employed in vehicle suspension design. The semi-active control device is similar to the passive devices in which properties of the damper can be adjusted such that spring stiffness and damping coefficient of the damper can be changed; thus, they are robustly stable. That is why the semi-active suspension system is widely used in the automotive industry.

S. Kashem et al., *Vehicle Suspension Systems and Electromagnetic Dampers*,
Springer Tracts in Mechanical Engineering, DOI 10.1007/978-981-10-5478-5_2

Since Karnopp et al. [52] developed the skyhook control strategy, extensive research has been done in semi-active control strategies [1–11]. Most of this research has been done to find practical and easy implementation methods or to achieve a higher level of vibration isolation or both. Adaptive-passive and semi-active vibration isolation is able to change the suspension system properties, such as spring stiffness and damping rate of the damper or actuator as a function of time. But the properties are changed relatively slowly in an adaptive-passive suspension system. However, in the semi-active system, the suspension properties are able to change within a cycle of vibration. The linear quadratic control is able to achieve both comfort and road holding improvements through the semi-active or active suspension system. But it requires the full-state measurement or estimation which is difficult to achieve [53, 54]. Linear time-invariant H-infinity control (LTIH) is able to provide better results, improving both ride comfort and road handling, ensuring predefined frequency behaviour [54]. Due to the fixed weights, this control system is limited to provide fixed performances [55, 56]. In 2006, Giorgetti et al. [57] compared different semi-active control strategies based on optimal control. They proposed a hybrid model with predictive optimal controller [54]. This control law is implemented via a hybrid controller, which is able to switch between large numbers of controllers that depend on the function of the prediction horizon [54]. It also requires a full-state measurement which is difficult to achieve. Recently, the uses of linear parameter-varying (LPV) approaches have become quite popular [54, 58, 59]. An LPV controller can either improve the robustness considering the non-linearities of the system or adapt the performances according to measured signals of road displacement and suspension deflection [54, 56, 60]. Another MPC system has been proposed by Canale et al. in 2006 [61]. The MPC controller is able to provide good performances, but it requires an online fast optimization procedure [54]. As it involves optimal control approach, a good knowledge of the model parameters and the full-state measurements is necessary to design the control system [54, 62]. Choudhury et al. [63] compared active and passive control strategies based on PID controller. There are many semi-active control systems designed, implemented and tested by many researchers. A few of them are described briefly in the following subsections.

2.1.1 Linear Quadratic Regulator and Linear Quadratic Gaussian

In the field of vehicle suspension control systems, the LQR approach is a widely used and studied control system. It has been studied and derived for a simple quarter-car model [64], half-vehicle model [65] and also full-vehicle model [66]. An optimal result is possible to achieve when the factors of the performance index such that acceleration of the body and dynamic tyre load variation are taken into account. In the LQR approach, a state estimator must be utilized if all the states are not available in the system, such as tyre deflections are difficult to measure in a moving vehicle. An estimator can narrow the phase margin of the LQR suspension system to a great extent, but it heightens the stability problems of the vehicle, especially if the suspension system is a fully active system. To solve this problem, Doyle and Stein proposed that the desired gain and phase properties can be obtained with a proper choice of estimator gains [67]. When implementing the LQR system on a full vehicle, another problem arises. The Riccati equation of the LQR system must be solved numerically for a full-vehicle model. The equation becomes very complex even though the vehicle is assumed to be symmetrical, and all the nonlinear effects created by the inertial effects and kinematical properties of the suspension system are not included. Different types of numerical algorithms are proposed to solve this issue, but none of them could guarantee convergence and the stability of the solution. The possibility of achieving a convergent solution decreases significantly when the number of actuator decreases or the order of the control system increases, or both, in the same system [68].

The LQR approach has also the inability to take the changes in steady state into consideration. These changes are caused by the change of payload at steady-state cornering of the vehicle. Elmadany and Abduljabbar [64] discussed a method to overcome this problem. That method is integral control. The task of integral control is to ensure the zero steady-state offset which would be applied to a quarter-car model. For a full-vehicle model, the integrator itself can deteriorate the performance of the controller. The proper selection of the integrator term and the gain of the integration time are a difficult problem in this approach due to the external forces caused by the non-zero offset which varies widely.

The optimal control method has been commonly used to accomplish a better comfort or handling the performance of a vehicle. Hrovat [69] has done extensive research with half-car models, full-car models, one degree of freedom models and two degrees of freedom models. He minimized the cost functions of the system combining excessive suspension stroke, sprung-mass jerk and sprung-mass acceleration together using linear quadratic (LQ) optimal control.

Shisheie et al. [70] presented a novel algorithm based on the LQR approach. It is able to optimally tune the PI controller's gains of a first order plus time-delay system. In this approach, the cost function's weighting matrices are adjusted by damping ratio and the natural frequency of the closed-loop system. In 1995, Prokop and Sharp [71] used LQR and LQG optimal control theories utilizing road preview data or information to get better ride quality. But the fact is, with respect to the system modelling errors, the LQG controller is less robust, and still today, determining the weighting coefficients for the LQG is a very hard job. According to Shen [72], most of the weighting coefficients for LQG/LQR control have been concluded by trial and error. Shen also revealed that the renowned skyhook feedback strategy provides the best outputs for the optimal feedback gain which reduces the mean square control effort and the cost function of the sprung mass' mean square velocity.

2.1.2 Sliding Mode Control

In the last 20 years, SMC has become one of the most active parts of control theory exploration. This exploration has established successful applications in a variety of engineering control systems, for example, aircraft, automotive engines, suspension, electrical motors and robot manipulators [73–75]. Shiri [76] has designed a sliding mode controller that is robust to electric resistance changes and bounded mass and also able to reject external disturbances. The simplicity system makes it adaptable to the electromagnetic suspension system. The results of the simulation confirm the robustness and the satisfactory performance of the designed controller against uncertainties and disturbances. There has also been a considerable amount of research done on the development of the theory of SMC problems for different types of systems, such as the fuzzy systems [77], the stochastic systems [78, 79] and the uncertain systems [80].

In a real dynamical system, it is impossible to avoid uncertainties due to the external disturbances and the modelling of the system. What is crucial is a solution to the robust control problem for uncertain systems. SMC can be used to deal with this problem. It is able to work with both uncertain linear and nonlinear systems successfully in a unified framework [81]. SMC design gives a systematic approach to the problem of maintaining consistent performance and stability in the face of the system's modelling imprecision. Since the variable structure with sliding mode (VSM) possesses the intrinsic nature of robustness, the VSM is found to be an effective technique to control the systems with uncertainties [82]. But the drawback of this system is when the system reaches the sliding mode state, the system with variable structure control becomes insensitive to the variations of the plant parameters. Many different techniques to design sliding mode controllers exist, but the baselines of all the techniques are very similar and can be divided into two main steps.

Firstly, design the control law of SMC in such a way that the trajectories of the closed-loop motion of the system are directed toward the SMC sliding surface, and make an effort to keep the motion on the surface thereafter.

Secondly, develop the sliding surface in the state space in such a way that the reduced-order sliding motion is able to satisfy the specifications specified by the designers.

Chan et al. [82] introduced a novel PID type SMC in which the sliding mode starts at the initial instant. As a result, during the entire process, the robustness of the system can be guaranteed. This system is also called an integral sliding mode control (ISMC). Yagiz et al. [83] has proposed and developed a sliding mode controller for a nonlinear vehicle model to overcome the problem of fault diagnosis and tolerance. A modified SMC was designed by Chamseddine et al. [84] for a linear full-vehicle active suspension system with partial knowledge of states of the system. For the conventional SMC strategy, the desired dynamic state can only be achieved when the sliding mode occurs.

2.1.3 Fuzzy and Neuro-Fuzzy Control

A vehicle suspension system is highly nonlinear and very complicated. Suspension actuation force changes when a vehicle rides on different road conditions. Conventional control strategies are not able to adapt to different environmental conditions. Fuzzy and neuro-fuzzy strategies can be used in controlled suspension systems in many ways. Fuzzy logic control (FLC) is appropriate for nonlinear systems. It can work with a complex system with no precise math model. This is why FLC is used in semi-active and active suspension systems to control the disturbance rejection. FLC is able to be insensitive to model and parameter inaccuracies with proper membership functions and rule bases.

To calculate the desired damping coefficients for semi-active systems, FLC can be utilized directly according to Al-Holou and Shaout [85]. Al-Holou and Shaout compared FLC to both passive and skyhook controllers. The authors employed FLC to the semi-active actuator to calculate the desired damping coefficient. In this study, a wide range of semi-active actuators was used. An important finding of this manuscript was that most of the FLC systems show similar results to the skyhook control system. It has been found that compared to the skyhook control system, a fuzzy-controlled semi-active suspension system showed slightly smaller RMS values of the body acceleration. Al-Holou and Shaout also showed that the semi-active suspension system with FLC increased the variation of dynamic tyre contact force compared to the skyhook-controlled semi-active suspension system.

FLC can also be used to calculate the required force for the active suspension system [86]. Barr and Ray compared the fuzzy-controlled active system with both the passive suspension system and the LQR active suspension systems. The authors have shown that the ride handling characteristic (the variation of dynamic tyre load) of FLC is better than the LQR and the passive suspension system. This result is slightly surprising, at least in the LQR active suspension system case. Moreover, the LQR-regulator cost function was not presented in this manuscript.

On the other hand, neural networks consist of a variety of alternative features such as computation, distributed representation, massive parallelism, adaptability, generalization ability and inherent contextual information processing. They can be utilized to model different types of ambiguities and uncertainties, which are often experienced in real life. Yan et al. [87] presented a multi-body vehicle dynamics model using ADAMS and a multilayer feedforward neural network of a series-parallel structure. The weights and threshold of neural networks have been optimized in this manuscript. The result of the combined simulation of MATLAB and ADAMS shows that the network convergence took place rapidly and the maximum error of identification is $<0.05\%$. The authors claimed that the designed genetic neural network can avoid the difficulty of establishing an accurately mathematical model for the vehicle semi-active suspension system.

The main objective of the hybridization of the control systems (using neural networks and fuzzy logic) is to overcome the weaknesses in one technology by using the strengths of the other during its application with appropriate integration.

In the majority of the studies concerning neural networks and fuzzy logic, the force of the actuator of the active suspension system or the damping coefficient of the semi-active suspension system is not controlled directly. Choi et al. [9] proposed a combination of neuro-fuzzy control approach to dictate a military-tracked vehicle semi-active suspension system. The fuzzification phase of the presented controller was continuously modified through a neural network. In this study, the models of real existing electrorheological semi-active actuator units and a 16 degrees of freedom vehicle model were utilized. For Direct Current Motor speed control online, Youssef et al. [88] have proposed an adaptive particle swarm optimization method for adapting the weights of fuzzy neural networks. An adaptive neuro-fuzzy control has been introduced by Khalid et al. [89] on the basis of particle swarm optimization-tuned subtractive clustering to provide critical information about the presence or absence of a fault in a two-tank process. Kashani and Strelow derived [90] a control system which consists of multiple LQG controllers around different operating points of the suspension system and blended the desired control actions of each controller with a fuzzy logic mixed algorithm. FLC was utilized to prevent the suspension from bottoming in this study. Kashani and Strelow claimed that this type of blending of a controller action is a fruitful idea and is able to improve the vehicle suspension system. But the limitations of practical implementation, such as maximum free rattle space, can be taken into account with decision logic of FLC.

2.1.4 Skyhook Control Method

The skyhook control is an effective vibration control algorithm which is able to dissipate the energy of the system at a high rate. For more than three decades, the skyhook control strategy has been widely researched. In 1974, Karnopp et al. [14] introduced the skyhook control strategy which is still used frequently in vehicle suspension applications. The name "skyhook" originates from the idea where a passive damper is imagined to be hooked from an imaginary inertial reference point or the sky. Skyhook damping is a damping force that is in the opposite direction to the sprung-mass absolute velocity and is proportional to the absolute velocity of the sprung mass (Fig. 2.1).

The above figure shows an ideal configuration of the skyhook semi-active control which has a sprung mass m_s hooked by a damper with skyhook damping constant c_{sky} from an imaginary sky (fixed ceiling); hence the name "skyhook" was used. If the damping force of the skyhook damper is F_{damp}, then the ideal skyhook control law can be expressed as:

$$\boldsymbol{F}_{\mathbf{damp}} = -\boldsymbol{c}_{\mathbf{sky}}\dot{\boldsymbol{x}}_{\mathbf{s}} \tag{2.1}$$

Here, x_s is the displacement. The skyhook-controlled semi-active suspension system (damper) utilizes a small amount of energy to run a valve, which adjusts the damping force. The damper valve can be a fluid valve or a mechanical element if it

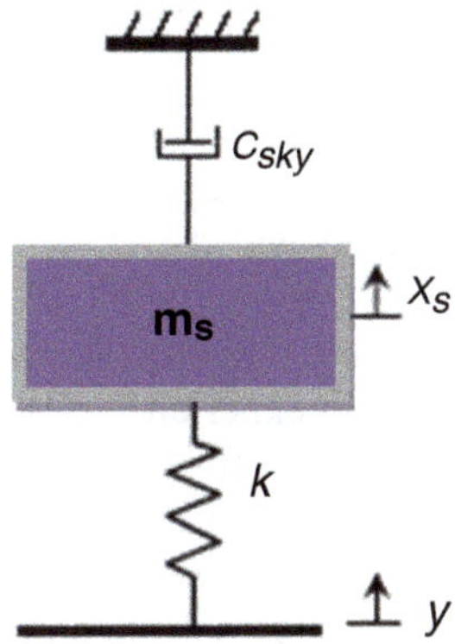

Fig. 2.1 An ideal skyhook configuration

is a mechanically adjustable damper. In a magnetorheological (MR) damper, the behaviour of rheological fluid changes according to the designed control system.

The active continuous skyhook control policy can also be ideally realized using an actuator or active force generator. Karnopp et al. [14] proposed the skyhook having a two-state control scheme named an ON–OFF control system. This control strategy switches between high and low damping states in order to achieve body comfort specifications [54]. But this control policy offers the damping force as equal to zero when the direction of sprung-mass velocity and the relative velocity of the sprung mass with respect to unsprung mass or ground is opposite. But in practice applying, zero damping force is not practicable for any semi-active damper. In 1974 Karnopp et al. [14] realized the complexity of the skyhook ON–OFF control method when it claims the force is needed to be equal to zero. However, because of the simplicity and practical implementation of the skyhook ON–OFF control strategy, it is widely used for vehicle suspension control [91]. In 1983, Karnopp [92] also proposed a new approach for a semi-active control system which consists of a variable stiffness method. In this control scheme, the damper is in a series connection with a spring of high stiffness, and the author suggested changing the stiffness of the spring according to the change in the damping coefficient of the damper.

Ahmadian and Vahdati [5] revealed that much research has been done on other variations of the skyhook control strategy in the past two decades, such as ON–OFF skyhook control, optimal skyhook control, continuous skyhook control and its modified versions. Li and Goodall [93] have introduced different control strategies which apply the skyhook damping control strategy for railway vehicle's active suspension system.

In 1983 Margolis [94] proposed another ON–OFF control method which simply switches off the damper when the unsprung mass and the sprung mass move in the same direction, and the unsprung mass has larger velocity than the sprung mass. Savaresi et al. proposed mixed skyhook and the ADD control approach [95, 96] which is a comfort-oriented control strategy having the switching strategy. Many researchers have investigated the clipped approaches which lead to unpredictable behaviours [57, 61]. Bessinger et al. [15] presented a modified skyhook control strategy. They modified the original skyhook control strategy proposed by Karnopp et al. in 1974 [14]. Bakar et al. [97] have also investigated the same strategy in their research. According to this modified skyhook control algorithm, both the passive damper and the skyhook damper effects are included to overcome the problem caused by the application of the original skyhook controller known as the water hammer [98, 99]. The water hammer problem is one in which the passengers of the vehicle experience unwanted audible noise and harsh jerks produced by the discontinuous forces (caused by low damping switches to high damping or vice versa). Nguyen et al. [51] have proposed a new semi-active control strategy called the optimal skyhook control approach. Soliman et al. [100] proposed an active suspension system controller employing the fuzzy-skyhook control strategy. This control system offered a new opportunity for vehicle ride performance improvement. The simulation result presented in the study shows the improvement of the vehicle ride quality by the proposed active suspension system with the fuzzy-skyhook control strategy. Compared to the passive suspension system, the body acceleration of the proposed system decreased. The suspension working space and the dynamic tyre load of the model show better performances too. Islam et al. [101] used skyhook control to compare the performance of magnetorheological, linear passive and asymmetric nonlinear dampers. Saad Kashem et al. [102] have proposed a new modified continuous skyhook control strategy with adaptive gain which dictates the vehicle's semi-active suspension system. The proposed closed-loop feedback system first captures the road profile input over a certain period. Then it calculates the best possible value of the skyhook gain for the subsequent process. Meanwhile, the system is controlled according to the new modified skyhook control law using an initial or previous value of the skyhook gain. In this chapter, the proposed suspension system is compared with passive and other recently reported skyhook-controlled semi-active suspension systems. Its performances have been evaluated in terms of ride comfort and road-handling performance. The model has been validated in accordance with the international standards of admissible acceleration levels ISO2631 and human vibration perception.

2.1.5 Groundhook Control Method

The groundhook control approach is almost similar to Karnopp's ON–OFF skyhook control method [14], except that the control system is based on the unsprung-mass damping control, as shown in Fig. 2.2.

The groundhook semi-active suspension system is a tyre displacement control system of a passive damper where one end is hooked on the ground or road surface and the other end is hooked to the tyre. The main idea of the groundhook control strategy is that it can be utilized to minimize the tyre contact force variation. These vibrational forces have a large impact on a vehicle's manoeuvrability and road-handling performance [103, 104]. Valášek et al. [105] have dealt with the novel

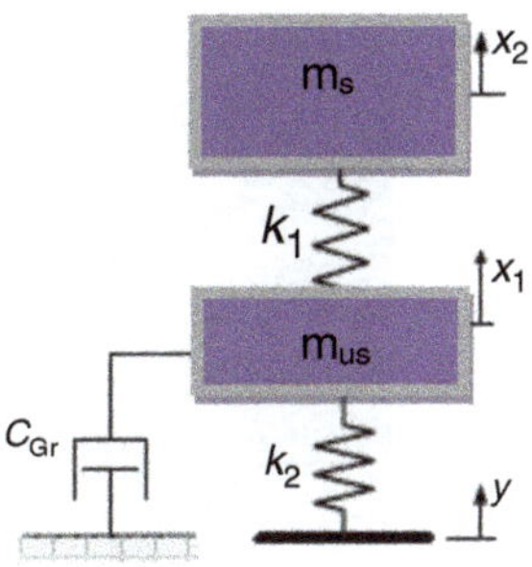

Fig. 2.2 A schematic of the groundhook control system

groundhook control concept for both active and semi-active suspension system of vehicles. Their ultimate objective is to reduce the tyre road forces of the suspension system. They have extended the basic groundhook control concept to several variants that enable the controller to increase driver comfort and decrease criteria of road damage for a broad range of road disturbances. The parameter optimization procedure has been used to determine the parameters of the control scheme for the generally nonlinear model. The influence and interaction of the time constants and damping rate limits of the variable shock absorbers are also addressed in this groundhook control approach.

2.2 Active Tilting Technology

The concept of "active tilting technology" has become quite popular in narrow tilting road vehicles and modern railway vehicles. Now in Europe, most new high-speed trains are fitted with active tilt control systems, and these trains are used as regional express trains [20, 21]. The description of tilting road vehicle technology is given in Sect. 2.2.1.

2.2.1 Narrow Titling Road Vehicle

Narrow vehicles are characterized by a high centre of gravity and relatively narrow track width compared to the standard production vehicle. These vehicles would be more efficient and pragmatic considering parking problems and traffic congestion in urban areas. They would also reduce energy consumption. These new cars are small, approximately half of the width of a conventional car (<2.5 m in length, 1 m in width and 1.5 m in height). All over the world, traffic congestion is a growing problem. Furthermore, the average number of occupants including the driver of a single vehicle in the USA is 1.57 persons (Fig. 2.3).

The narrow commuter vehicle can be categorised into two types depending on their tiling mechanisms. The first one, Fig. 2.4a, uses an active suspension system to

Fig. 2.3 Narrow commuter vehicle [106]

(a) **(b)**

Fig. 2.4 (**a**) Vehicle tilt by suspension [107], (**b**) vehicle tilt by actuator [108]

tilt the whole vehicle, and the second one, Fig. 2.4b, has an actively controlled tilting passenger cabin and a non-tilting chassis frame or rear assembly. An actuator fitted to the rear assembly controls the tilt action of the passenger cabin according to the design criteria. The non-tilting assembly of the vehicle typically consists of several powertrain components, so therefore it contributes considerably to the mass and inertia of the vehicle. Moreover, the non-tilting chassis has to support the roll torque which has been applied to tilt the passenger cabin by the actuator. As a result, the suspension of the vehicle wheel needs to be quite stiff which may affect the ride comfort. Furthermore, the energy consumption of this tilting mechanism is also very high.

This particular small and narrow geometric property of the vehicle poses stability problems while cornering or lane change. There are also two types of control schemes that have been used to stabilize the narrow tilting vehicle [31]. These control schemes are defined as direct tilt control (DTC) and steering tilt control

(STC) systems as detailed in [27, 32, 109]. In the DTC system, the driver steering input is connected to the front wheel steering mechanism directly [31]. In a DTC system, dedicated actuators control the tilt of the vehicle (such as having an active suspension). In this system, the link between the wheels and the steering wheel is no longer mechanical. In an STC system, on the other hand, STC or steering tilt control, no additional actuator is used, and the tilt of the vehicle is controlled by the steering angle input from the driver. The steering input is used to follow the desired trajectory as well as stabilize the tilt mode of the vehicle. This is particularly a steer-by-wire system [31]. In this system, the driver steering input signal is read by the controller, and the controller determines the tilt angle. Since the beginning of the 1950s, extensive research has been done on both types of control systems by the automotive industry and researchers.

Motorized tilting vehicles have been studied and developed since the pioneering prototype proposed by Amati et al. [43] in 1945–1950. The Ford Motor Company developed a two-wheeled lean vehicle in the middle of the 1950s [43]. It was gyroscopically stabilized with retractable wheel pods for parking [43]. In the 1960s, the MIT presented a tilting vehicle which was equipped with an active roll control [43]. The design was similar to a motorcycle. At the beginning of the 1970s, General Motors developed a tilting vehicle called the "Lean Machine". It had a fixed rear frame and a tilting body module that were controlled by the rider. The rider had to balance the tilting body using foot pedals [27, 43].

More recently, Brink Dynamics [25] developed a three-wheeled car named Carver with a rotating body and non-tilting rear engine. BMW and the Universities of Bath and Berlin presented Clever in 2003 [110]. It consists of a non-tilting two-wheel rear axle and a single front wheel that tilts with the main body. The rear body remains in contact with the ground in the same way as a conventional automobile rear axle, but the main body is connected to the rear frame by a suspension layout enabling it to lean like a motorcycle.

The manufacturer Lumeneo presented the Smera and Piaggio MP3 [111]. At the Tokyo motor show 2009, Nissan revealed the Land Glider [22], which is a four-wheeled narrow vehicle. Of all the above, the Carver One was sold commercially between 2006 and mid-2009, and the MP3 has been on the market for sale since 2006 [43].

From an academic point of view, researchers have done an extensive amount of work on these cars. D. Karnopp suggested that the narrow tilting vehicle would have to lean into a corner and also explained the optimum desired lean angle in his research [26]. Dean Karnopp and his co-workers have also carried out a significant amount of research into dynamic modelling of tilting vehicles [31]. Karnopp and Hibbard have proposed that a tilt actuator can be employed to tilt a narrow tilting vehicle to a certain desired tilt angle with the help of the direct tilt control strategy [26]. It is apparent that their research lays down the basic ideas for designing a direct tilt control system. However, in some of their research [26–28], they are unable to take into account the lateral position acceleration of the vehicle while calculating the desired tilt angle calculation. This caused the controller to require a high transient torque.

There are a few publications which have presented the idea of a virtual driver in a narrow tilting vehicle. These virtual drivers are able to follow a path without falling to one side. Saccon et al. [29] developed a dynamic inversion of a simplified motorcycle model. This model is able to obtain a stabilizing feedback through the standard linear quadratic regulatory control system. This model allows the controller to calculate the state and input trajectories according to a desired output trajectory of the tilting vehicle. To avoid the direct deal with the lean instability, Frezza and Beghi [30] took the roll angle as control input instead of the steering angle input from the driver. They have defined the path tracking as an optimization problem of the controller design.

Snell [112] proposed to start the tilting action with the STC system then to switch to the DTC system to maintain the tilting position. A three-wheeled prototype of a narrow tilting vehicle was developed at the University of Bath, UK. It employed hydraulic actuators to tilt the cabin with the help of DTC technology which has a high power requirement [113]. Kidane et al. [114] applied hybrid control schemes with both STC and DTC. This work employed a feedforward plus PID controllers to stabilize the tilt of the vehicle, and a look-ahead error of the trajectory model was used as the driver model. Chiou proposed a double-loop PID to control and to maintain the tilting position and the rate of the vehicle [115].

Defoort [116] and Nenner et al. [117] worked with the trajectory-tracking and robust stabilization problems of a riderless bicycle. They developed a dynamic model that considers geometric-stabilization mechanisms. They also derived a combined control system consisting of a second-order sliding mode controller and disturbance observer. In their research, they adopted a simplified tricycle model as the dynamic model of a bicycle.

In addition, in the research works stated above and in other authors' researches, the effect of road bank angle is not considered in the control system design and in the modelling of the dynamic model of narrow tilting vehicles [26, 27, 32, 36–44]. The result of not incorporating road bank angle is a non-zero steady-state torque requirement. It also significantly increases transient torque requirements. So and Karnopp [28] considered the road bank angle in their work, but it has no effect on the final form of the control input [31]. The authors specified that the lateral acceleration of the vehicle is obtained from the sensor readings mounted on the vehicle. But it is evident that the reading of an accelerometer of a narrow tilting vehicle would be contaminated by the tilt angle, the road bank angle and the angular acceleration of the vehicle [31].

2.2.2 *Tilting Standard Production Vehicle*

To improve vehicle performance during cornering or sudden lane change, advanced electromechanical and electronic systems are used, for example, antilock braking systems, electronic brake force distribution, active steering and electronic stability programs. Nowadays, some researchers have focused on active steering control to improve vehicle cornering [17–19]. Recently, a system was presented by Bose Corporation, namely, the Bose suspension system [34]. This system consists of a power amplifier and a linear electromagnetic motor at each wheel that is controlled by a set of control algorithms. The high-bandwidth linear electromagnetic dampers of this system respond quickly enough to achieve better ride performance. To date, the prototype of the Bose suspension system is installed in standard production vehicles and able to achieve superior comfort and control simultaneously. According to the manufacturer, the Bose suspension system can counter the body roll of the vehicle by stiffening the suspension while cornering. It can also change the ride height dynamically and is capable of performing the four quadrant operations and the high-bandwidth operation. But it uses less than one-third of the power of the air conditioning system of a typical vehicle. However, to date, no commercial tests or design details are available to the world from the Bose Corporation which would allow an accurate and unbiased comparison with other competitive suspension systems.

Vehicle performance during cornering has been improved by most car manufacturers using electronic stability control (ESC). Car manufacturers use different brand names for ESC: Volvo calls it dynamic stability and traction control (DSTC), Mercedes and Holden call it electronic stability program (ESP) and BMW and Jaguar call it dynamic stability control (DSC), but whatever the term used, the processes are almost same. To avoid oversteering and understeering during cornering, ESC extends the brake and different torque on each wheel of the vehicle. But ESC reduces the longevity of the tyre because of the tyre skids during random braking. To overcome this problem, a vehicle can be tilted inward via an active or semi-active suspension system.

Car giant Nissan has developed a four-wheeled ground vehicle for the future which is half scooter and half car [35]. The electric-powered Land Glider shown in Fig. 2.5 is approximately half the width of a family car and is designed for busy city streets. It uses a steer-by-wire system to control the vehicle manoeuvre and has small motors mounted on each wheel. A computer in the Land Glider automatically calculates the amount of lean required to corner considering the speed, steering

Fig. 2.5 Nissan Land Glider [35]

angle and yaw rate of the vehicle. The vehicle body can lean into a corner up to 17° for sharper handling. In addition, in the works stated above and other authors' researches, the effect of road bank angle is considered neither in the control system design nor in the modelling of the dynamic model of the tilting vehicles.

2.3 Conclusion

For a long time, active and semi-active suspension systems have been employed as a practical application of modern control theory. In this literature review, many robust and optimal control approaches or algorithms have been reviewed including linear time-invariant H-infinity control (LTIH), linear parameter-varying control (LPV) and MPC. Five widely known control approaches are reviewed more deeply, namely, the LQR and LQG, SMC, fuzzy and neuro-fuzzy control and the skyhook and groundhook approaches. It has been found that the skyhook control strategy is the most widely used due to its simplicity for practical implementation. But still, there is a great scope of work yet to be done to modify the skyhook control strategy to achieve better performance. Different types of damper technologies have also been discussed in this chapter, and it has been shown that the linear electromagnetic damper is best for the semi-active suspension system due to its fast response time which is better than the best hydraulic device. A brief literature review on automotive tilting technology has also been done in this chapter. This highlights that a direct tilting method needs to be developed to tilt the standard passenger vehicle inward during cornering while considering the road bank angle.

Chapter 3
Vehicle Suspension System

Abstract The quarter-car suspension model is the best benchmark to study and analyse the dynamic behaviour of vehicle vertical isolation properties. This chapter presents background information and a description of the quarter-car suspension model which includes passive, semi-active and active suspension. This chapter also consists of a comparison of various models to determine the appropriate quarter-car model to compare the control systems discussed in Chap. 4.

3.1 Vehicle Suspension System

A suspension system is an essential element of a vehicle to isolate the frame of the vehicle from road disturbances. Figure 3.1 shown here is a typical car suspension system. It is required to maintain continuous contact between a vehicle's tyres and the road. The most important element of a suspension system is the damper. It reduces the consequences of an unexpected bump on the road by smoothing out the shock. In most shock absorbers, vibration energy is converted to heat and dissipates into the environment. In the viscous damper, energy is converted to heat via a viscous fluid. In hydraulic cylinders, the hydraulic fluid is heated up. In air cylinders, the hot air is emitted into the atmosphere. But the electromagnetic damper is different; here the vibration energy is converted into electricity via an electric motor (induction machine or DC motor or synchronous machine) and stored in a condenser or battery for further use [119].

Suspension systems are categorized as passive, active and semi-active considering their level of controllability. Although all the types of the suspension systems have different advantages and disadvantages, all of them utilize the spring and damper units.

3.1.1 Passive Suspension System

Passive suspension systems are composed of conventional springs and oil dampers with constant damping properties (Fig. 3.2). In this model m_1 and m_2 represent the

S. Kashem et al., *Vehicle Suspension Systems and Electromagnetic Dampers*,
Springer Tracts in Mechanical Engineering, DOI 10.1007/978-981-10-5478-5_3

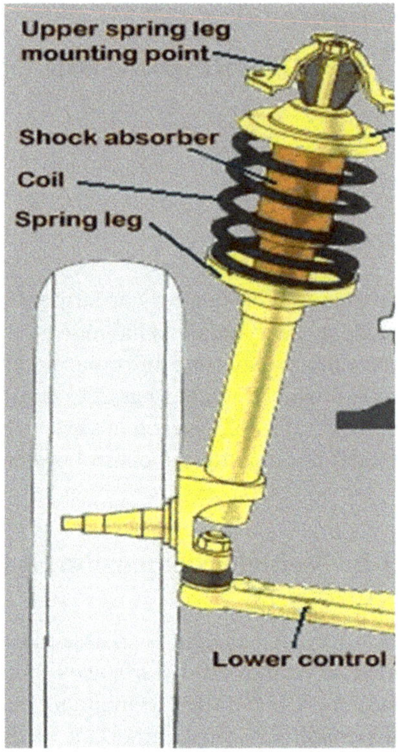

Fig. 3.1 Suspension system [118]

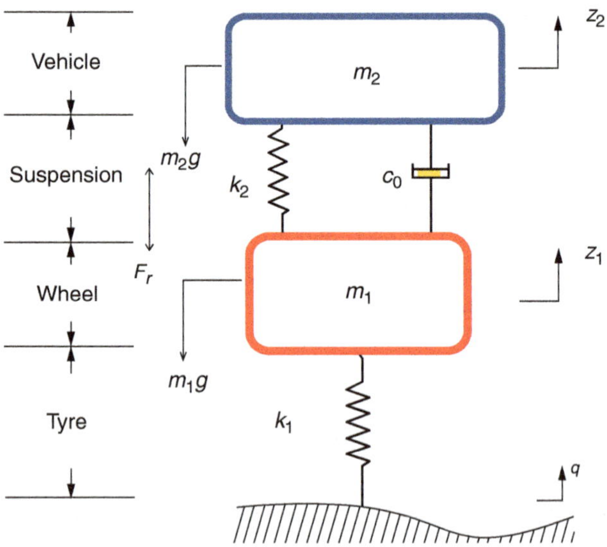

Fig. 3.2 Passive suspension system

unsprung mass and sprung mass, respectively, k_1 is the tyre stiffness coefficient or tyre spring constant and k_2 is the suspension stiffness or suspension spring constant. c_0 and c_t are the suspension damping constant and the tyre damping constant, respectively, F_r is friction of suspension and q, z_1 and z_2 represent road profile input, displacement of unsprung mass and displacement of sprung mass, respectively

In most instances, passive suspension systems are less complex, more reliable and less costly compared to active or semi-active suspension systems. The constant damping characteristic is the main disadvantage of passive suspension systems. For a passive suspension, the use of soft springs and moderate to low damping rates is needed, but the use of stiff springs and high damping rates is needed to reduce the effects of dynamic forces. Designers utilize soft springs and a damper with low damping rates for applications that need a smooth and comfortable ride such as in a luxury automobile.

On the other hand, sports cars incorporate stiff springs and a damper with high damping rates to gain greater stability and control at the expense of comfort. Therefore, the performance in each area is limited to the two opposing goals [120]. There is always a compensation need to be made between ride comfort and ride handling in the passive suspension system as spring and damper characteristics cannot be changed according to the road profile.

3.1.2 Semi-active Suspension System

The semi-active suspension system was first proposed by Karnopp et al. in 1973 [52]. Figure 3.3 shows a semi-active suspension model. Here f_d can generate an active actuating force by an intelligent controller. Since then, semi-active suspension systems have continued to acquire popularity in vehicular suspension system applications, due to their better performance and advantageous characteristics over passive suspension systems. In semi-active suspension systems, the damping properties of the damper can be changed to some extent. The adjustable damping characteristics in semi-active dampers are achieved through a variety of technologies, such as electrorheological (ER) and magnetorheological (MR) fluids, solenoid valves and piezoelectric actuators. It has been widely recognized that a semi-active suspension system provides better performance than a passive system. As it is safe, economical and does not need a large power supply, the semi-active suspension has recently been commercialized for use in high-performance automobiles [121–125]. However, there still exist many challenges that have to be overcome for these technologies to achieve their full potential. MR degradation with time, sealing problems and temperature sensitivity are some crucial issues of the MR dampers that need development.

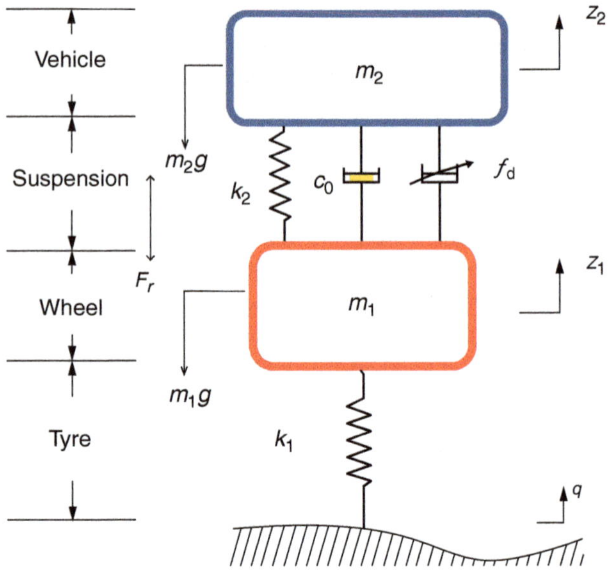

Fig. 3.3 Semi-active suspension system

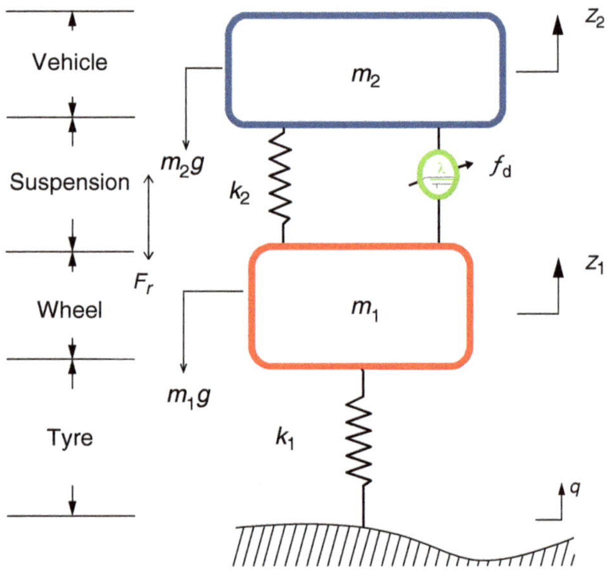

Fig. 3.4 Active suspension system

3.1.3 Active Suspension System

The active suspension system (Fig. 3.4) actuates the suspension system links by extending or contracting them through an active power source as required [120]. Conventionally, automotive suspension designs have been a compromise between the three contradictory criteria of road handling, suspension travel and passengers' comfort. In recent years, the use of active suspension systems has allowed car manufacturers to achieve all three desired criteria independently. A similar approach has also been used in train bogies to improve the curving behaviour of the trains and decrease the acceleration perceived by passengers. But this makes the system expensive and increases the design complexity and energy demands.

From the above discussion, it is apparent that a semi-active suspension system is more appropriate for implementing and evaluating the performance of various control strategies [120–125].

3.2 Quarter-Car Suspension Model

In this manuscript, two degrees of freedom quarter-car model has been used to evaluate the performance of various controllers as described in Chap. 4. A quarter-car model imitates the heave or the vertical motion of the vehicle alone. As the design goal of the most semi-active suspension system is to reduce the vertical acceleration, the quarter-car model is sufficient for evaluating the performance of control strategies [53]. The sprung mass, suspension components, unsprung mass and a wheel are the basic components of a quarter-car model. For a quarter-car model, sprung mass means the body or chassis of the car, and it represents almost one-fourth of the weight of the whole body of the car. The suspension system bridges the connection between the wheel and body of the car and consists of many parts and varies according to the type of the suspension system such as passive, semi-active or active suspension (described in the previous section). Unsprung mass includes the weight of everything geometrically below the suspension system, such as an axle, wheel and rim. The wheel denotes the tyre, which incorporates the spring and damping characteristics.

Two degrees of freedom quarter-car model as shown in Fig. 3.5a is known as an ideal model and used by some researchers [126–128]. Faheem et al. [129] presented an

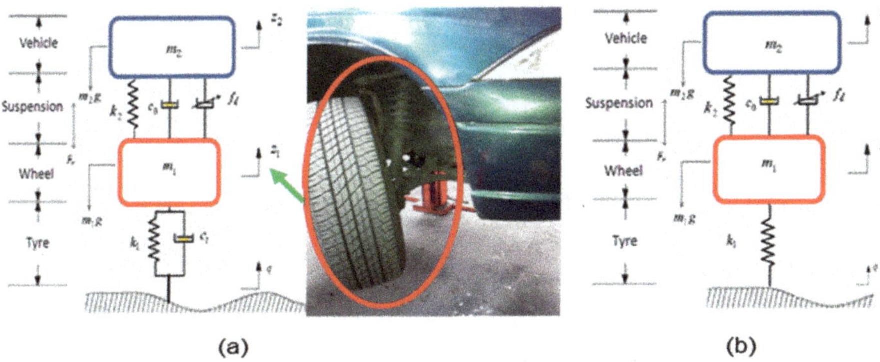

Fig. 3.5 (**a**) Ideal quarter-car model, (**b**) simplified quarter-car model

insight on the suspension dynamics of the quarter-car model with a complete state-space realization. In the ideal case, the sprung mass and unsprung mass are free only to bounce vertically. The ideal dynamic equations of motion of unsprung and sprung masses which satisfy Newton's second law of motion are given by Eq. (3.1):

$$\begin{aligned} m_1\ddot{z}_1 &= -c_0\left(\dot{z}_1-\dot{z}_2\right)-k_2(z_1-z_2)-c_t\left(\dot{z}_1-\dot{q}\right)-k_1(z_1-q)+f_d-F_r+m_1g \\ m_2\ddot{z}_2 &= -c_0\left(\dot{z}_2-\dot{z}_1\right)-k_2(z_2-z_1)-f_d+F_r+m_2g \end{aligned} \tag{3.1}$$

The simplified model as shown in (b) has been used in most recent studies [16, 130–133] as the effect of the tyre damping coefficient c_t is negligible compared to the tyre stiffness coefficient. So omitting the tyre damping force $c_t\,(\dot{z}_2-\dot{q})$, Eq. (3.1) becomes:

$$\begin{aligned} m_1\ddot{z}_1 &= -c_0\left(\dot{z}_1-\dot{z}_2\right)-k_2(z_1-z_2)-k_1(z_1-q)+f_d-F_r+m_1g \\ m_2\ddot{z}_2 &= -c_0\left(\dot{z}_2-\dot{z}_1\right)-k_2(z_2-z_1)-f_d+F_r+m_2g \end{aligned} \tag{3.2}$$

3.2.1 Explanation of Motion Equations of Quarter-Car

To understand the motion equations for the quarter-car suspension, it is better to start from ideal mass-spring-damper motion equations, which are well known. The first one considers the horizontal motion as shown in Fig. 3.6.

In this figure, x is the position of the square block in meters, m is the mass of the block in kilograms, k is the spring stiffness in Newton's per meter and F_{spring} is the spring force in Newton's. When a spring is stretched from its equilibrium position due to an external force, the spring itself acts as a force proportional to the length it is stretched, and this force acts in the opposite direction to the stretch:

$$F_{\text{spring}} \infty - \text{stretch}$$

Or

$$F_{\text{spring}} = -k \times \text{stretch}$$

If $x=0$ at the position where the spring is in equilibrium, then x is equal to the stretch of the spring. So the force of the spring becomes:

$$F_{\text{spring}} = -k\,x$$

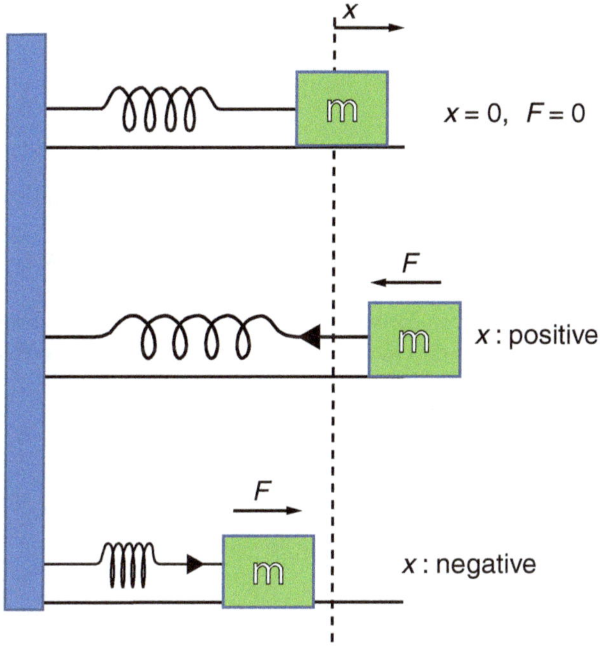

Fig. 3.6 Mass spring characteristics

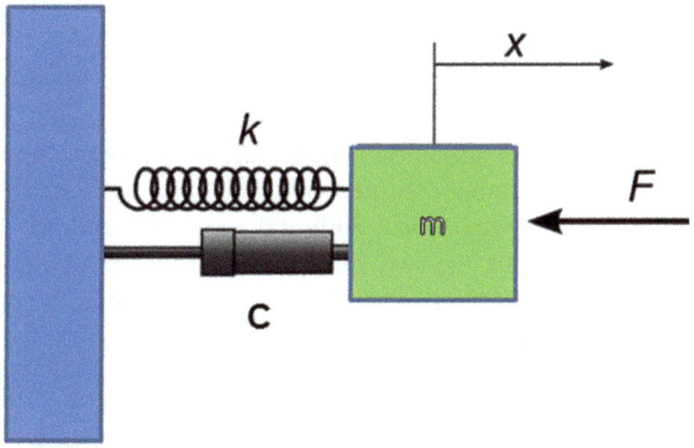

Fig. 3.7 Mass-spring-damper configuration

In addition, there is a force that opposes the motion of the mass as shown in Fig. 3.7.

In this figure, c is the damping constant in Newton-second per meter and v is the velocity of the block in meters per second. This force is the damping force and it is proportional to the mass velocity which also opposes the mass velocity, such as:

$$F_{\text{damping}} \propto -v$$

Or

$$F_{\text{damping}} = -c\,v$$

So the total force acting on spring-mass-damping system is:

$$F = F_{\text{spring}} + F_{\text{damping}} = -k\,x - c\,v$$

According to Newton's law of motion, $F = ma$. From the definition of acceleration, the first derivative of position x is equal to the velocity v and the acceleration a is equal to the second derivative of position x. So:

$$a = \ddot{x}$$

and

$$v = \dot{x}$$

Now the differential equation becomes:

$$m\ddot{x} = -c\dot{x} - kx \tag{3.3}$$

The simple mass-spring-damper model described above is the foundation of vibration analysis. This is defined as the single degree of freedom (SDOF) model, since it has been assumed that the mass only moves up and down in the same axis. Figure 3.8 is a more complex system involving more mass which is free to move in more than one direction – adding degrees of freedom.

In this model, the two springs act independently, so it is easy to figure out the forces acting on the two blocks. It is assumed that the connection of the spring and

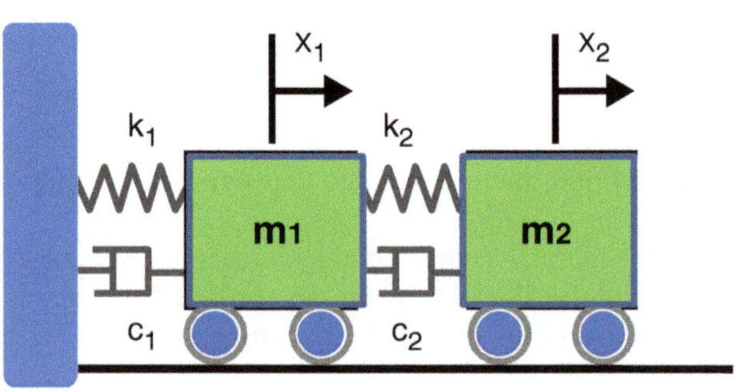

Fig. 3.8 Two degrees of freedom horizontal multiple mass spring damper

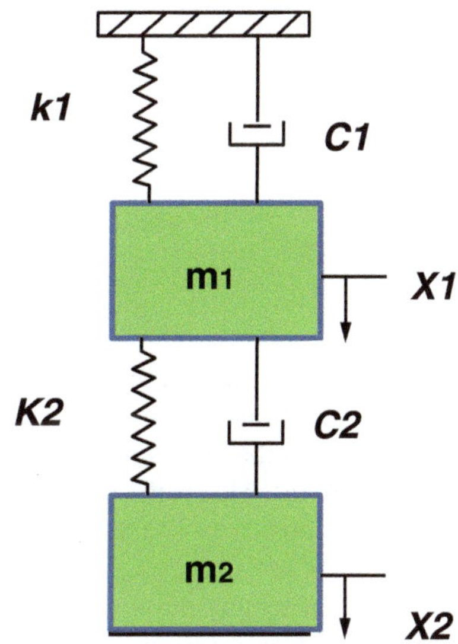

Fig. 3.9 Vertical multiple mass spring-damper configuration

damper to the wall is the origin of this suspended system. Here x_1, x_2 are the position (left edge) of the blocks, m_1, m_2 are the mass of blocks and k_1, k_2 are the spring constants. So the motion equations would be (Fig. 3.9):

$$\begin{aligned} m_1 \ddot{x}_1 &= -c_2\left(\dot{x}_1 - \dot{x}_2\right) - c_1 \dot{x}_1 - k_2(x_1 - x_2) - k_1 x_1 \\ m_2 \ddot{x}_2 &= -c_2\left(\dot{x}_2 - \dot{x}_1\right) - k_2(x_2 - x_1) \end{aligned} \tag{3.4}$$

Now the vertical linear motion has been considered as shown in the above figure. Here a new force strikes due to gravitation g (m/s^2) which acts in the same direction (downward) as the mass velocity and equals the product of mass and gravity, so the differential equation becomes:

$$\begin{aligned} m_1 \ddot{x}_1 &= -c_2\left(\dot{x}_1 - \dot{x}_2\right) - c_1 \dot{x}_1 - k_2(x_1 - x_2) - k_1 x_1 - m_1 g \\ m_2 \ddot{x}_2 &= -c_2\left(\dot{x}_2 - \dot{x}_1\right) - k_2(x_2 - x_1) - m_2 g \end{aligned} \tag{3.5}$$

Now, consider two degrees of freedom quarter-car suspension model having an actuator which delivers a force f_{d} as shown in (a) and the corresponding motion equation.

If one considers the forces acting on the unsprung mass m_1, then the forces acting downward is the $m_1 g$ force due to gravitation and actuating force f_{d}. As shown in

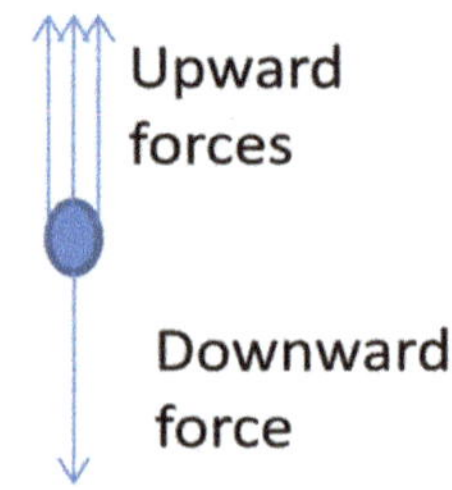

Fig. 3.10 Forces acting at a point

Fig. 3.10, force due to the acceleration of the unsprung mass $m_1 \ddot{z}_1$ is acting in the upward direction. If the displacement $z_1 > q$ is positive, then the spring force $k_1(z_1 - q)$ and the damping force $c_1(z_1 - q)$ are negative in the downward direction as shown in (a). This is the same for a damping force of co and a spring force of k_2 if $z_1 > z_2$ is positive. The friction force F_r is acting negatively in the downward direction.

Again for sprung mass m_2, the forces acting downward are the $m_1 g$ force due to gravity and friction force F_r. The force due to the acceleration of the sprung mass $m_2 \ddot{z}_2$ is acting in the upward direction. The actuating force f_d is acting negatively downward. The damping force of co and spring force of k_2 is negative in the downward direction under the condition that displacement $z_2 > z_1$.

3.2.2 High- Versus Low-Bandwidth Suspension System

A semi-active suspension system has two sections: semi-active and passive. The semi-active part usually gets damping force from an external energy source to control the suspension system (in the regenerative type system, it may differ). The passive part has a spring and a damper or similar devices. In some systems, this part is rigid, but it can be omitted as well. This can be distinguished as low-bandwidth and high-bandwidth suspension systems [134].

Low-bandwidth configuration (LBC) represents the series connection between the active and passive components of the suspension system (Fig. 3.11a). In the mathematical modelling, the differential motion equations are as follows:

$$\begin{aligned} m_1 \ddot{z}_1 &= -c_0\left(\dot{z}_{1b} - \dot{z}_2\right) - k_2(z_{1b} - z_2) - k_1(z_1 - q) - F_r + m_1 g \\ m_2 \ddot{z}_2 &= -c_0\left(\dot{z}_2 - \dot{z}_{1b}\right) - k_2(z_2 - z_{1b}) + F_r + m_2 g \end{aligned} \tag{3.6}$$

1

where $c(dz_{1b} - dz_1) = f_d$ is the actuating force. Through LBC configuration, the active suspension system can control the car body (sprung mass) height. But the

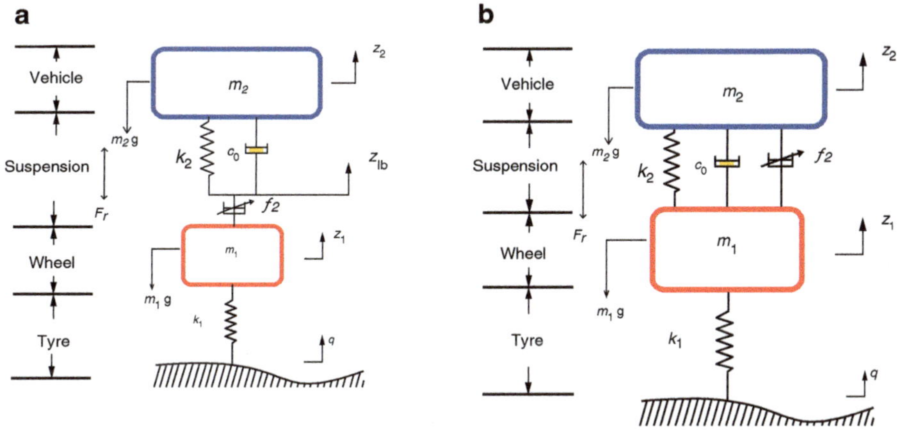

Fig. 3.11 (**a**) Low-bandwidth suspension model, (**b**) high-bandwidth suspension model

actuator cannot be omitted or turned off as it carries the static load. Another disadvantage of this system is that it is good only in the low-frequency range.

On the other hand, high-bandwidth configuration (HBC) is possible to control at higher frequencies than low-bandwidth configuration (LBC), and also the passive part can work alone in case of failure of the active part. The only drawback of HBC is that practically it can't control the vehicle height. In an HBC configuration, active and passive components are linked in parallel (Fig. 3.11b). The motion equations of HBC are almost similar to that of LBC, but an extra term is added which is an actuator force f_d.

In this manuscript, two degrees of freedom HBC semi-active suspension system is used, mainly because there is no requirement of a static load force.

3.3 Comparison of Recent Models

Comparison between different quarter-car suspension systems described in the literature has been done considering the quarter-car model travelling over a ramp in the street as shown in Fig. 3.12. This is the road profile input q for the quarter-car suspension model.

The quarter-car suspension model used in this comparison analysis has been shown in (b). The parameters of quarter-car models chosen by different authors in their recent research have been extracted and described in the table below.

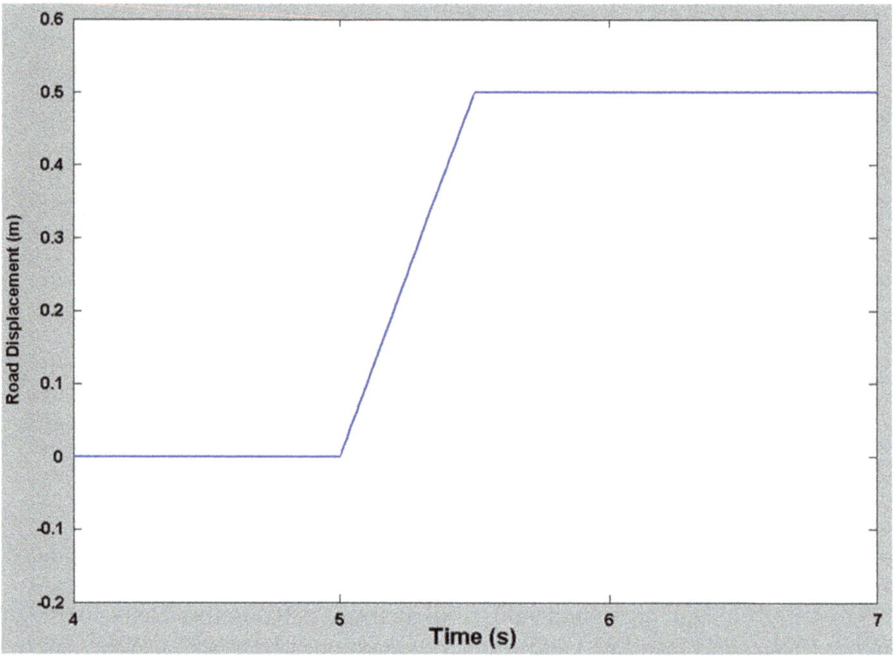

Fig. 3.12 The road profile

The above-stated models have been simulated in the MATLAB/Simulink environment, and the sprung-mass acceleration of quarter-car models of different authors has been shown in the figures below.

As the negative amplitudes are almost the opposite mirror of the positive ones, these are not shown in the figure above.

Figure 3.13 shows the comparison between the different authors of quarter-car models based on their chosen parameters, m_1, m_2, k_1, k_2 and c_2, as stated in Table 3.1. In this comparison peak amplitude and settling time of sprung-mass acceleration have been considered.

As shown in Table 3.2, the model used by Zhang et al. [126] has a peak amplitude reduced by 17% from Chen's [1] model, whereas the settling time increases by 23%. Again, in a comparison with Chen's [1] model, the model used by Priyandoko et al. [132] has a settling time that decreases by 5%, but peak amplitude increases by 6%.

Furthermore, the overall performance of Chen's [1] model is the best compromise compared to other models in terms of peak amplitude and settling time.

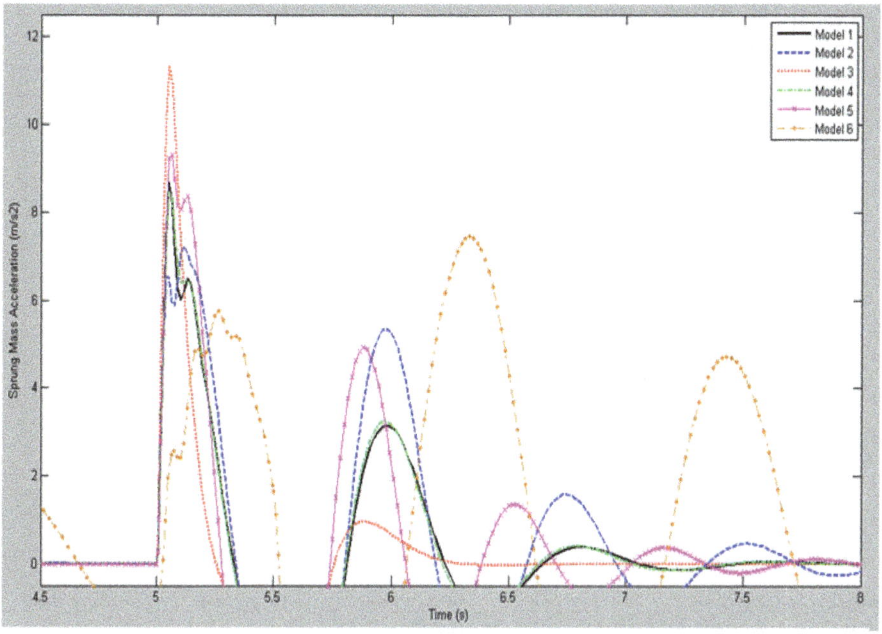

(a)

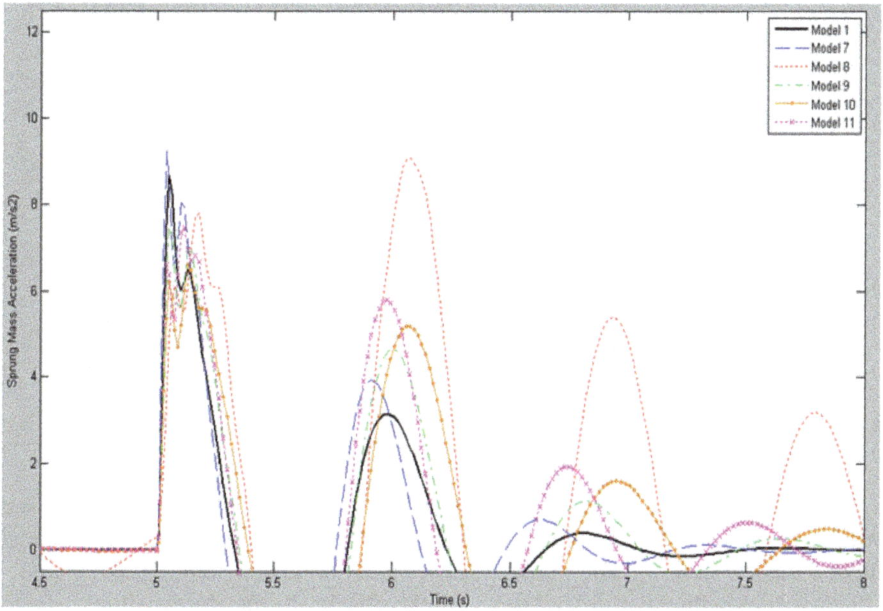

(b)

Fig. 3.13 (**a**) Comparison between passive suspension models 1–6, (**b**) comparison between passive suspension models 1 and 7–11

Table 3.1 The parameters of quarter-car models

Model No.	Model Authors	m_1 (kg)	m_2 (kg)	k_1 (N/m)	k_2 (N/m)	c_0 (N-s/m)
1	Chen [1]	36	240	160,000	16,000	1400
2	Zhang et al. [126]	31	380	228,000	29,000	1500
3	Jia-ling and Jia-qiang [135]	20	160	100,000	10,000	2000
4	Nguyen and Choi [51]	31	230	130,000	16,000	1400
5	Scheibe and Smith [133]	35	250	150,000	29,000	1400
6	Gopala Rao and Narayanan [136]	1009	1000	360,000	36,000	1000
7	Priyandoko et al. [132]	25	180	190,000	16,000	1000
8	Fateh and Alavi [128]	40	243	124,660	14,671	370
9	Gao et al. [2]	36	240	160,000	16,000	980
10	Du et al. [137]	40	320	200,000	18,000	1000
11	Sung et al. [138]	40	380	309,511	28,516	1284

Table 3.2 Comparison between outputs of the vehicle sprung-mass acceleration

Model No.	Model Authors	Maximum positive amplitude (m/s^2)	Settling time (s)
1	Chen [1]	8.69	2.87
2	Zhang et al. [126]	7.24	3.53
3	Jia-ling and Jia-qiang [135]	11.31	2.33
4	Nguyen and Choi [51]	8.47	2.98
5	Scheibe and Smith [133]	9.31	2.97
6	Gopala Rao and Narayanan [136]	5.77	12.4
7	Priyandoko et al. [132]	9.23	2.72
8	Fateh and Alavi [128]	7.11	11.4
9	Gao et al. [2]	7.49	3.65
10	Du et al. [137]	6.61	4.12
11	Sung et al. [138]	7.46	3.33

3.4 Conclusions

In this chapter, the vehicle suspension system has been categorized and discussed briefly. It has been explained that the semi-active suspension system is the most suitable for the investigation in this manuscript. A brief description of the quarter-car model has been given as well as an explanation of the motion equations used in the model. High- and low-bandwidth suspension systems have also been discussed. As there is no requirement of a static load force in this manuscript, in Chap. 4, two degrees of freedom HBC semi-active suspension system is used to investigate different semi-active control algorithms. An extensive comparison of quarter-car models extracted from the study of various researchers has been presented in this chapter. From the comparison, it has been concluded that the better estimation of the parameters of mass, spring constant and damping coefficient for a passive quarter-car model is Chen's [1] model, and this model has been used for further investigation. The semi-active control algorithms experiments are discussed in the next chapter.

Chapter 4
Design of Semi-active Suspension System

Abstract One of the key issues in the design of an active or semi-active suspension system is to identify the appropriate control algorithm. This chapter describes the proposed modified skyhook control closed-loop feedback system and its effectiveness in a semi-active suspension system. The chapter comprises four main sections. The first section describes the proposed and three existing skyhook control algorithms, while the second section describes the road profile that needs to be generated to evaluate the controller performances. The third section presents the simulation of the quarter-car model as described in Chap. 3 with the semi-active control algorithms. The last section is comprised of simulation and experimental analysis of the Quanser quarter-car suspension plant designed and manufactured by Quanser Inc. The last two sections also compare the results of different control techniques and evaluate the proposed modified skyhook control algorithm. The comparison has been done in terms of ride comfort and road-handling performance. On the other hand, the evaluation consists of a human vibration perception test and admissible acceleration levels test based on ISO 2631.

4.1 Semi-active Control Algorithms

On the basis of the two degrees of freedom semi-active suspension system described in Chap. 3, passive and four semi-active suspension systems have been modelled. The continuous skyhook control of Karnopp et al. [14], modified skyhook control of Bessinger et al. [15], optimal skyhook control of Nguyen et al. [51] and the proposed modified skyhook control strategies are used in designing the semi-active suspension system. The control strategies are described below.

4.1.1 Continuous Skyhook Control of Karnopp et al. [14]

The semi-active continuous skyhook control strategy of Karnopp et al. [14] can be represented by the following equation:

S. Kashem et al., *Vehicle Suspension Systems and Electromagnetic Dampers*,
Springer Tracts in Mechanical Engineering, DOI 10.1007/978-981-10-5478-5_4

$$f_d = \begin{cases} \left[\max\left[C_{\min}.\min\left[\frac{C_{\text{sky}}\dot{z}_2}{\left|\dot{z}_2-\dot{z}_1\right|}C_{\max}\right]\right]\right] \times \left|\dot{z}_2-\dot{z}_1\right| & \text{for } \dot{z}_2\left|\dot{z}_2-\dot{z}_1\right| \geq 0 \\ C_{\min} \times \left(\dot{z}_2-\dot{z}_1\right) & \text{for } \dot{z}_2\left|\dot{z}_2-\dot{z}_1\right| < 0 \end{cases} \tag{4.1}$$

where f_{d} is the semi-active damping force of the actuator.

This strategy is used in many recent studies [139, 140]. According to this control strategy, the effective damping of the skyhook damper is bounded by a high and a low level. Determining whether the damper is to be adjusted to either its low state or its high state depends on the product of the velocity of the spring mass attached to that damper $\dot{z}_2$ and the relative velocity across the suspension damper $\dot{z}_2 - \dot{z}_1$. If this product is greater than or equal to zero, then the high state of the damper is applied. If this product is negative, the damper is adjusted to its low state. In this situation, it is better to supply no force at all, but in practice, the semi-active damper coefficient is limited by the physical parameters of the conventional damper, which means that there is both an upper bound, $C_{\max}$, and a lower bound, $C_{\min}$, and they have certain values depending on the chosen damper. Here C_{sky} is the nominal damping coefficient selected by the designer and $C_{\max} < C_{\text{sky}} < C_{\min}$.

4.1.2 Modified Skyhook Control of Bessinger et al. [15]

The modified skyhook control strategy presented by Bessinger et al. [15] is a modification of the original skyhook control strategy proposed by Karnopp et al. in 1974 [14]. Bakar et al. [97] have also used the same strategy in their research. Both the passive damper and skyhook damper effects are included in the modified skyhook control algorithm to overcome the problem caused by the application of the original skyhook controller known as the water hammer [98, 99]. The water hammer problem is one where the passenger experiences unwanted harsh jerks and audible noise created by the force discontinuity (caused by low damping switches to high damping or vice versa). The equation of the modified skyhook control algorithm is given by:

$$f_{\text{d}} = C_{\text{sky}}\left[\alpha\left(\dot{z}_2 - \dot{z}_1\right)\right] + (1-\alpha)\,\dot{z}_2 \tag{4.2}$$

where α is the passive to skyhook ratio and C_{sky} is the damping constant of a modified skyhook control. The value of α is chosen to be 0.5, and an optimal value of C_{sky} is chosen such that the desired force estimated from this control algorithm is to be within the range of damping forces of the designed damper.

4.1.3 Optimal Skyhook Control of Nguyen et al. [51]

Nguyen et al. [51] have described the semi-active optimal skyhook control strategy by the following equation:

$$f_{\text{actual}} = \begin{cases} f_{\max} & \text{if} \quad f_{\max} \le u \\ u & \text{if} \quad f_{\min} < u < f_{\max} \\ f_{\min} & \text{if} \quad f_{\min} \ge u \end{cases} \tag{4.3}$$

where $f_{\text{actual}} = f_{\text{d}}$ is the semi-active damping force of the actuator and $f_{\max} = C_{\max} \times (\dot{z}_2 - \dot{z}_1)$ and $f_{\min} = C_{\min} \times (\dot{z}_2 - \dot{z}_1)$ are the maximum and minimum damping forces that can be exerted by the actual damper at a given relative velocity, respectively. $u = C_{\text{sky}} \times \dot{z}_2$ is the damping force exerted by the damper where C_{sky} is the optimal damping coefficient obtained by the following equation in the simulation environment:

$$\left(C_{\text{sky}}\right)_{\text{Optimal}} = \min\{\text{RMS}\,[J]\} \tag{4.4}$$

$$\text{Where } \mathrm{RMS}[J] = \mathrm{RMS}\left[w_1\, \ddot{z}_2 + w_2(z_2 - z_1) + w_3(z_1 - q)\right] \tag{4.5}$$

Here w_i, $i = 1, 2, 3$ are the set values of weighting factors according to the defined objective which is minimizing the criterion of $C_{\text{sky (optimal)}}$.

4.1.4 Proposed Skyhook Control with Adaptive Skyhook Gain

The proposed modified skyhook control algorithm is chosen to provide the desired force in attenuating road harshness in the real world. From the discussion in the literature review and Sects. 4.1.1, 4.1.2 and 4.1.3, it has been understood that the designers have chosen or derived (trial and error methodology) a constant value of C_{sky} for their skyhook control strategy and used the same value for all road conditions. In the real world, a vehicle is not always travelling on the same type of road, and it is very difficult to imitate all types of road surfaces in a simulation environment. The road disturbance interacting with the vehicle tyre in the real world is quite different compared to the road disturbance modelled in the simulation environment. A good semi-active suspension system should provide high damping on good roads for better body isolation, low damping on average roads to achieve good comfort and finally, adequate damping on the poor roads for structural modes (road profile description is given in Sect. 4.2). Skyhook gain should be varied according to the road surface on which the vehicle is travelling. The proposed modified skyhook control strategy with adaptive gain (illustrated in

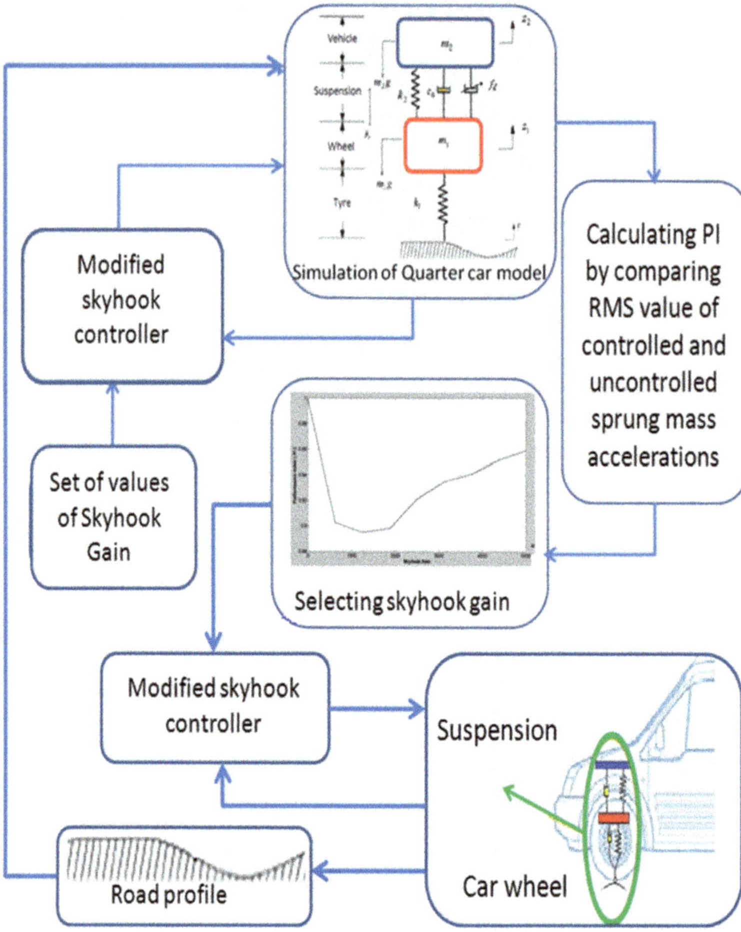

Fig. 4.1 Schematic of the suspension systems based on proposed modified skyhook control system with adaptive skyhook gain

Fig. 4.1) is developed to address this problem. A modification of conventional continuous skyhook control has been proposed and it is described by Eq. (4.6).

$$f_{\mathrm{d}} = \begin{cases} C_{\max}\left(\dot{z}_1 - \dot{z}_2\right) & \text{if } \dfrac{\dot{z}_2}{\left(\dot{z}_1 - \dot{z}_2\right)} - \dfrac{C_{\mathrm{sky}}}{C_{\max}} \\ C_{\mathrm{sky}}\,\dot{z}_2 & \text{if } \dfrac{C_{\mathrm{sky}}}{C_{\max}} \dfrac{\dot{z}_z}{\left(\dot{z}_1 - \dot{z}_2\right)} - \dfrac{C_{\mathrm{sky}}}{C_{\min}} \\ C_{\min}\left(\dot{z}_1 - \dot{z}_2\right) & \text{Otherwise} \end{cases} \tag{4.6}$$

C_{max} and C_{min} represent the maximum and minimum damping coefficient of the actuator, respectively. The value of C_{sky} is varied in accordance with the road profile input. The velocity of the sprung mass relative to the unsprung mass $(\dot{z}_2 - \dot{z}_1)$ is denoted as positive when the base and mass of the suspension system are splitting (i.e. when $z_2 > z_1$). If the sprung and unsprung masses are splitting, the semi-active damper becomes in tension. As a result, the force f_d works in the negative z_2 direction which is applied to the sprung mass by the actuator. In the following equation, f_d is pointing in the opposite direction of z_2:

$$f_d = -C_0 \times \dot{z}_2 \tag{4.7}$$

where C_0 is the required damping coefficient of the actuator. Since the actuator is capable of producing a force in the appropriate direction, the only requirement to match the skyhook suspension is:

$$C_0 = C_{sky} \frac{\left(\dot{z}_1 - \dot{z}_2\right)}{\dot{z}_2} \tag{4.8}$$

This control strategy dictates the actuator movement and also determines the specific value of C_{sky} for the road surface on which the vehicle is travelling. At first, the road profile input is captured by the tyre deflection measurement over a certain period of time while the vehicle is travelling on the road. Then the quarter-car model identical to one fourth of the vehicle suspension system is simulated in the simulation environment (onboard system placed in the vehicle) as both uncontrolled (passive) and controlled (semi-active) suspension systems. The controlled suspension system is dictated by the modified skyhook control strategy described in Eq. (4.6) with a range of C_{sky} (the range is depicted by the rated maximum and minimum damping coefficient of the actuator). The RMS values of sprung-mass acceleration of both controlled and uncontrolled suspension systems are calculated and a performance index (PI) is derived by the following equation described by Sung et al. [138]:

$$\frac{\mathrm{PI} = \sqrt{\sum_{i=1}^{N} \ddot{z}_{2,C}(i)^2}}{\sqrt{\sum_{i=1}^{N} \ddot{z}_{2,UC}(i)^2}} \tag{4.9}$$

Then the optimal value of C_{sky} is chosen for which the PI becomes minimum:

$$\left(C_{sky}\right)_{Optimal} = \min\{[\mathrm{PI}]\} \tag{4.10}$$

This optimal value of C_{sky} replaces the initial value of the skyhook gain of the modified skyhook controller which is dictating one fourth of the vehicle suspension system for the next certain period of time. This time, the interval would be determined by the processor speed of the onboard computer of a vehicle. While C_{sky} is calculated, the suspension system behaves according to the modified skyhook control law with an initial or previous value of C_{sky}. After each certain period of time interval, C_{sky} is adapted according to the road surface to achieve better performance. The whole process is represented in Fig. 4.1.

To investigate how the controller behaves under different road conditions, a road profile needs to be generated. The next section describes the road profile.

4.2 Road Profile Description

In order to validate the quarter-car model and its parameter estimation, one needs to run the simulation of the model on a road surface which is adopted by most researchers. To predict the response of quarter-car models, it is often beneficial to handle the excitation created by road irregularities as the spectrum of a geometrical road profile, $P(n)$ [141]. This can be expressed as:

$$P(n) = \begin{cases} P(n_0)\left(\frac{n}{n_0}\right)^{-w_1}, & \text{if } n \le n_0, \\ P(n_0)\left(\frac{n}{n_0}\right)^{-w_2}, & \text{if } n > n_0, \end{cases} \tag{4.11}$$

where $P(n_0)$ is the road roughness, n is a spatial frequency, $n_0 = 1/2\pi$ c/m and $w_1 = 2.0$ and $w_2 = 1.5$. The drop in magnitude is modelled by the waviness ω. ISO standards suggest $\omega = 2$ for road undulations [142]. In this study, three classes of roads are used as defined by ISO8608 [50] and their $P(n_0)$ values are given in Table 4.1.

The road irregularity also depends on vehicle speed. If a vehicle is travelling on L length of road segment with speed V, then the road profile is described by a superposition of $N \to \infty$ sine waves:

$$\mathrm{RP}(t) = \sum_{i=1}^{N} A_i \sin\left(\Omega_i t - \alpha_i\right) \tag{4.12}$$

where $A_i = \sqrt{(2p\ (\Omega_i)\ 2\pi/L)}$, $i = 1, 2, 3, N$, α_i is a random variable with a uniform distribution in the interval $[0, 2\pi]$ [43]. The value of Ω is determined by:

$$\Omega = \frac{2\pi}{L} V \tag{4.13}$$

Table 4.1 Theoretical road classes on the basis of road roughness

Road class	Road roughness $P(n_0)$ (m^3/c)
A (good)	16×10^{-6}
B (average)	64×10^{-6}
C (poor)	256×10^{-6}

It has been assumed that the vehicle speed $V = 20$ m/s, the distance travelled by the vehicle $L = 200$ m and the time taken $t = 10$ s. Superposition of $N = 100$ has been chosen in this analysis as the standard of ISO8606. Figure 4.2a and b describe the time histories and power spectral density of three classes of roads, respectively.

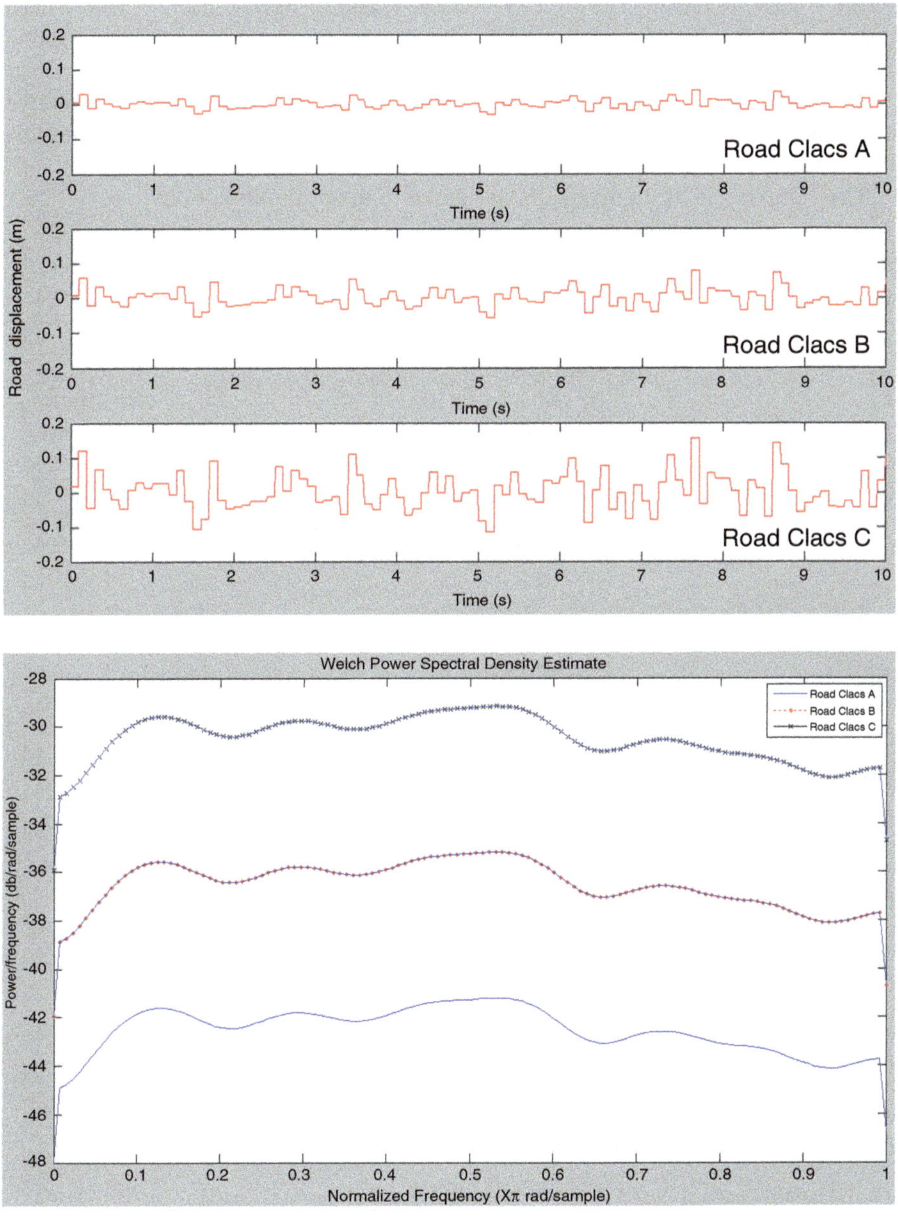

Fig. 4.2 (**a**) The time histories of three classes of roads, (**b**) power spectral density of three classes of road

4.3 Comparison and Evaluation Using Y. Chens' Model

The parameters of the two degrees of freedom of Chen's [1] quarter-car model used in Chap. 3 are listed in the following table (Table 4.2).

To investigate how the controller behaves under different road conditions, a road profile is generated using the three classes (A, B and C) of road profiles where each class lasts for the 20 s. Figure 4.3 shows the road profile input that has been used in the simulations to compare the performance of the controllers.

The reaction of the road profile input shown by the passive and four semi-active suspension systems described above is depicted in Fig. 4.4. Random 5 s along the total length of the simulation is shown for better visualization.

The comparison of the results is discussed in the next subsections.

Table 4.2 Nominal parameter values used in the simulation

Parameter	Definition	Values
m_1	Unsprung mass	240 (kg)
m_2	Sprung mass	36 (kg)
k_1	Tyre spring coefficient	160,000 (N/m)
k_2	Suspension spring coefficient	16,000 (N/m)
c_0	Suspension damping coefficient	1400 (N-s/m)
F_r	The friction of suspension	300 (N)
g	Gravity acceleration	9.8 (m/s^2)

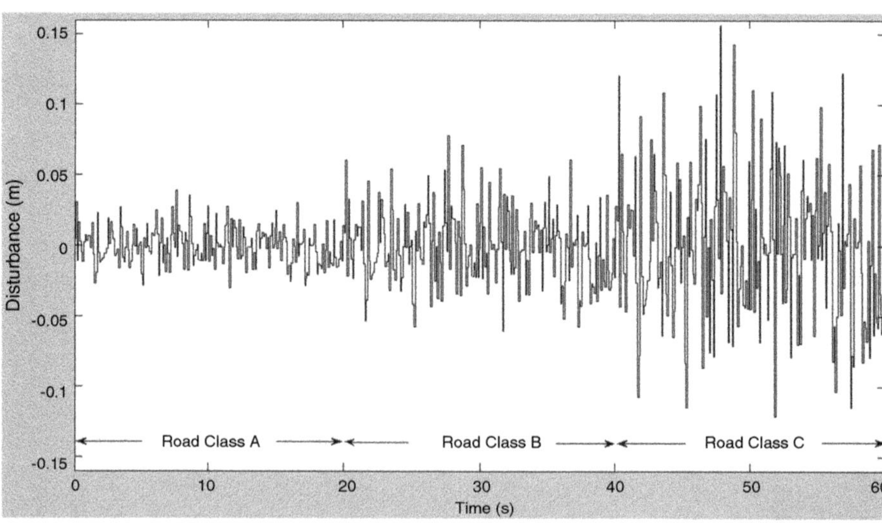

Fig. 4.3 The time history of road profile

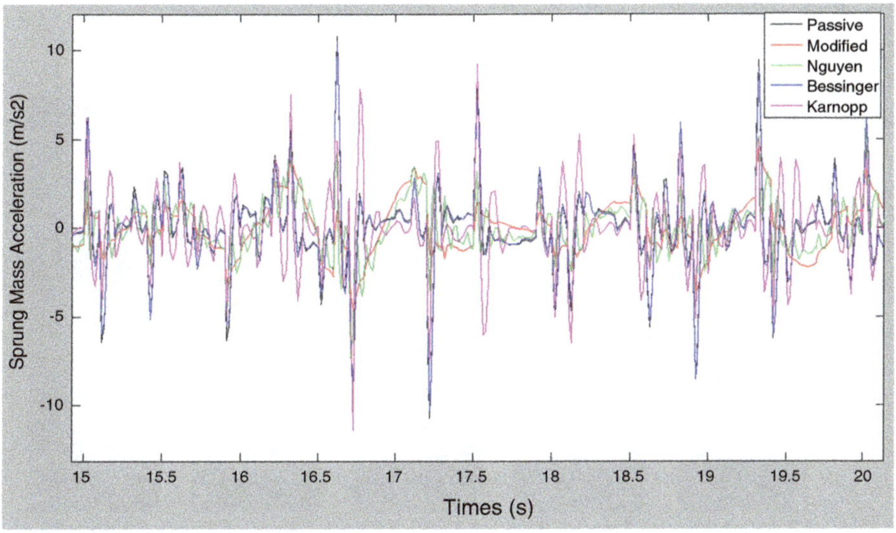

Fig. 4.4 The sprung-mass acceleration of the passive and semi-active suspension systems

4.3.1 Comparison

To compare, the controllers' performances, ride comfort and road-handling performance are calculated in the following section.

4.3.1.1 Comparison of Ride Comforts Performance

Ride comfort depends on human perceptions of vehicle noise, vibration and motion. While it varies from person to person, a quantitative evaluation of the ride comfort performance (RCP) can be carried out following the methodology provided by Abramov et al. [142] and that is the root mean square of the sprung-mass acceleration normalized by the gravity acceleration g:

$$\mathrm{NRMS} = \left(\frac{1}{T} \int_{t=0}^{T} \left[\frac{\ddot{z}_2\,(t)}{g} \right]^2 \mathrm{d}t \right)^{1/2} \tag{4.14}$$

Here T is the duration of exposure. Lower NRMS value represents higher passenger comfort [142]. The NRMS achieved in simulation and experiment by the passive and four semi-active suspension systems are shown in Fig. 4.5.

The figure above shows the improvements on the comfort index (NRMS) of the semi-active systems with respect to the passive suspension system. The proposed

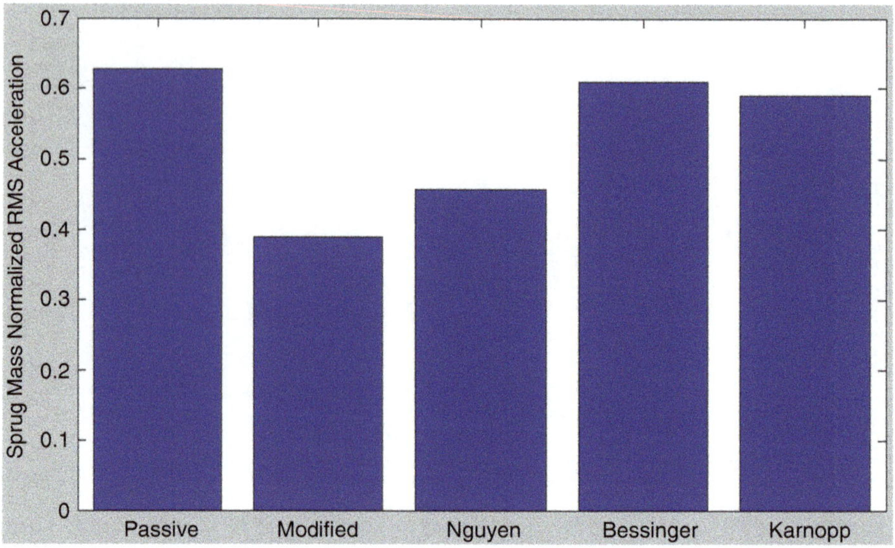

Fig. 4.5 The ride comfort performance comparison

modified skyhook control strategy, optimal skyhook control of Nguyen et al. [51], modified skyhook control of Bessinger et al. [15] and continuous skyhook control of Karnopp et al. [14] are improved by 38.4%, 27.3%, 2.8% and 5.9%, respectively.

4.3.1.2 Comparison on Road Handling Performance

To examine the ride-handling performance of a vehicle, the forces acting on the tyres and the road should be considered. The vehicle's reaction point with the road disturbance is the tyres. For a quarter-car model, the force exchanged between the road and tyre is $Fz_1 = k_1\,(z_1 - q)$. Then the vehicle's road-handling performance (RHP) can be calculated by the root mean square of the forces normalized by the static forces which act on the wheels. The static forces are represented as $(m_1 + m_2)g$ [142] (Fig. 4.6):

$$\mathrm{RHP} = \left(\frac{1}{T}\int\limits_{t=0}^{T}\left[\frac{F_{Z1}(t)}{(m_1 + m_2)g}\right]^2 \mathrm{d}t\right)^{1/2} \tag{4.15}$$

Higher RHP value represents higher ride-handling performance [142]. Compared to the passive suspension system, the figure above shows that the RHP of the semi-active suspension system controlled by the proposed modified skyhook control approach, optimal skyhook control of Nguyen et al. [51], modified skyhook control

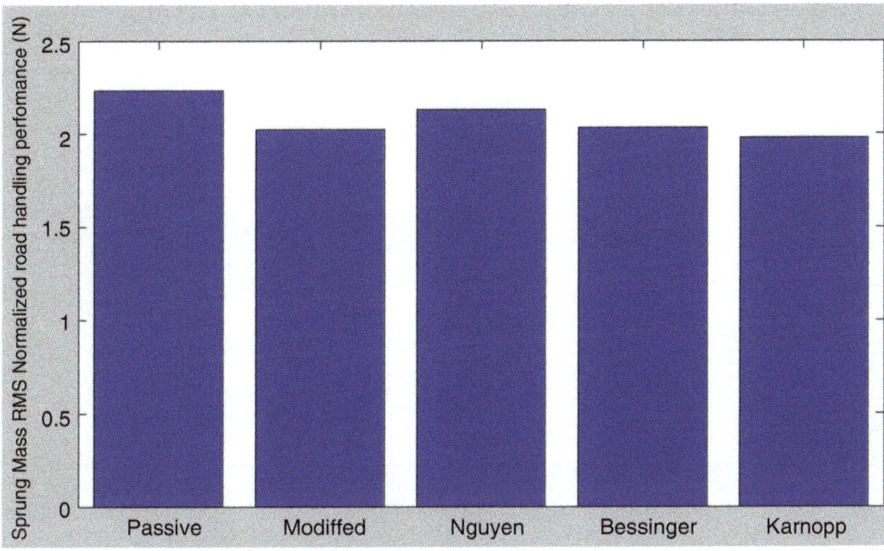

Fig. 4.6 The road-handling performance comparison

of Bessinger et al. [15] and continuous skyhook control of Karnopp et al. [14] decreases by 9.3%, 4.6%, 8.9% and 11.4%, respectively.

4.3.2 Evaluation

The proposed modified skyhook controller performance is validated in terms of the human vibration sensitivity test, and the ride comfort level test specified by the International Organization for Standardization (ISO 2631) [143].

4.3.2.1 Human Vibration Sensitivity

The human vibration sensitivity test is more suited for a full-car model instead of a quarter-car model, but if the quarter-car model is considered as a one-wheel vehicle or, in any instance, if the vehicle's body vibration depends only on one wheel, then this sensitivity test can be run on a quarter-car model [134]. The study shows that human beings are very sensitive to the vertical motion in the frequency range of 4–8 Hz [134]. So while designing a vehicle suspension, sprung-mass acceleration tolerance should be kept to a minimum, keeping in mind the frequency range mentioned above (Fig. 4.7).

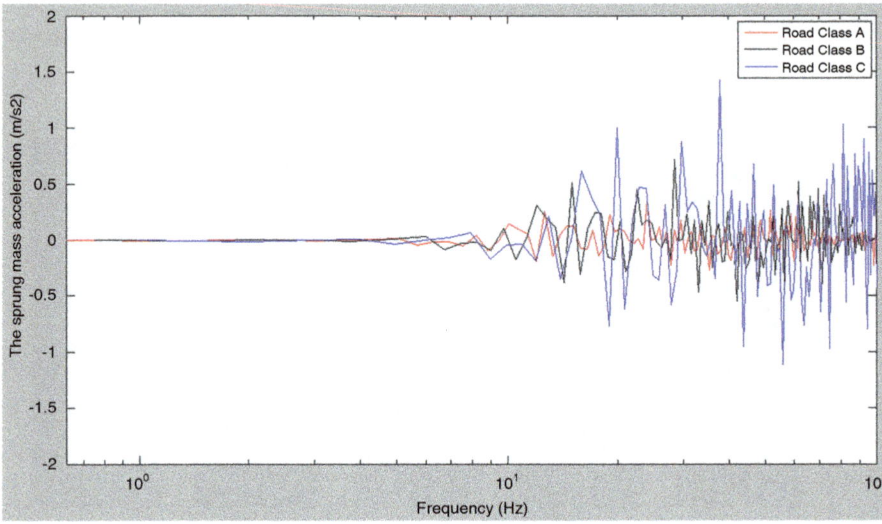

Fig. 4.7 Human vibration sensitivity test in frequency domain

In the figure above, it is clearly shown that at low frequency, vertical vibration of the sprung mass of the semi-active suspension system controlled by the proposed algorithm remains almost zero for each class of road disturbance.

4.3.2.2 Admissible Acceleration Levels Test Based on ISO 2631

Ride comfort depends on human sensitivity to vehicle noise, vibration and motion. It varies from person to person, but a quantitative evaluation of the ride comfort performance can be done following the methodology provided by ISO. The international code ISO 2631 [143] defined a term named weighted RMS acceleration (denoted here as WRMS) which can be formulated by the following equation:

$$\text{WRMS} = \left(\frac{1}{T} \int_{t=0}^{T} \sum_{i=1}^{N} \left[F_i \ddot{z}_{2,i}(t) \right]^2 \mathrm{d}t \right)^{1/2} \tag{4.16}$$

where $i = 1, 2, 3, N$, T is the duration of exposure and F_i is the weighting coefficient of acceleration in the vertical direction.

The WRMS value of the sprung mass obtained in this simulation is 0.71 for road class C. This confirms the validation of the proposed modified skyhook control system in terms of ride comfort enhancement for two degrees of freedom semi-active suspension system.

4.4 Comparison and Evaluation of Quanser Suspension Plant

In this study, two degrees of freedom suspension plant provided by Quanser [144] has been used to evaluate the controller's performance. The description of the plant is given below.

4.4.1 Quanser Quarter-Car Suspension Plant

The Quanser suspension plant is a bench-scale model which imitates two degrees of freedom quarter-car model dictated by a semi-active suspension system. Figure 4.8 shows the full model of the Quanser suspension plant.

As a nomenclature of the Quanser suspension system, the table below provides a list of all the principal elements combined in the equipment. On the Quanser suspension system represented in the figure above, each and every element is located and identified through a unique identification (ID) number in Fig. 4.9 (Table 4.3).

The plant has three plates on top of each other. The vehicle body is represented by the top plate and is suspended over the middle plate with two springs. An accelerometer is fitted at the top floor to measure the acceleration of the vehicle body with respect to the plant ground. Between the top and middle plates, a capstan drive DC motor is placed to emulate a semi-active or active suspension mechanism. The middle plate and the bottom plate represent the wheel assembly and the road exciter, respectively. These plates are connected through a spring-damper mechanism which emulates the tyre in the quarter-car model. The bottom plate is attached to a fast response DC motor which allows the designer to generate different road profiles.

The three plates can easily slide along a stainless steel shaft using linear bearings. A torque is generated at the output shaft when the DC motor turns. This torque is converted to a linear force which results in the bottom plate's motion through the lead screw and gearing mechanism. The motion of the top plate relative to the middle one is tracked directly by high-resolution optical encoders, while two other encoders measure the motion of the two bottom plates. This quarter-car structure has been used to study the semi-active control algorithms in this manuscript.

4.4.1.1 State-Space Representation

This Quanser suspension system can be modelled as a quarter-car model as (a) to simulate the real system. A state-space representation of the Quanser quarter-car model needs to be derived in order to simulate the real system and test control

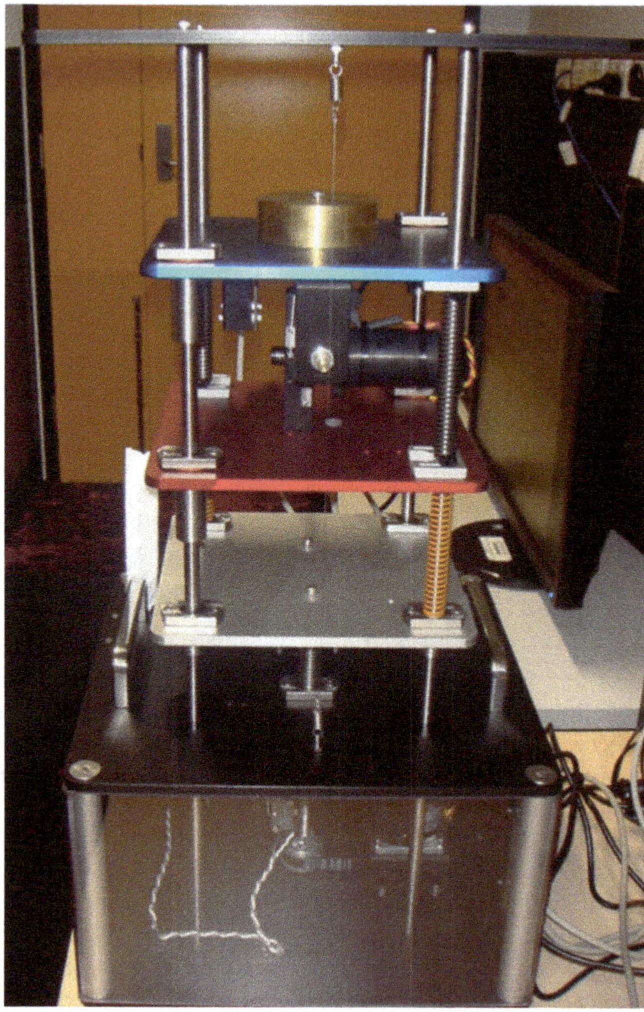

Fig. 4.8 Quanser suspension plant

strategies. In this section, a state-space representation of the Quanser suspension system will be derived. By definition, state-space matrices represent a set of linear differential equations that describe the system's dynamics. The two equations of motion of the Quanser quarter-car model are linear and time invariant and they can be represented as follows:

$$\begin{aligned} \dot{x} &= Ax + Bu \\ y &= Cx + Du \end{aligned} \tag{4.17}$$

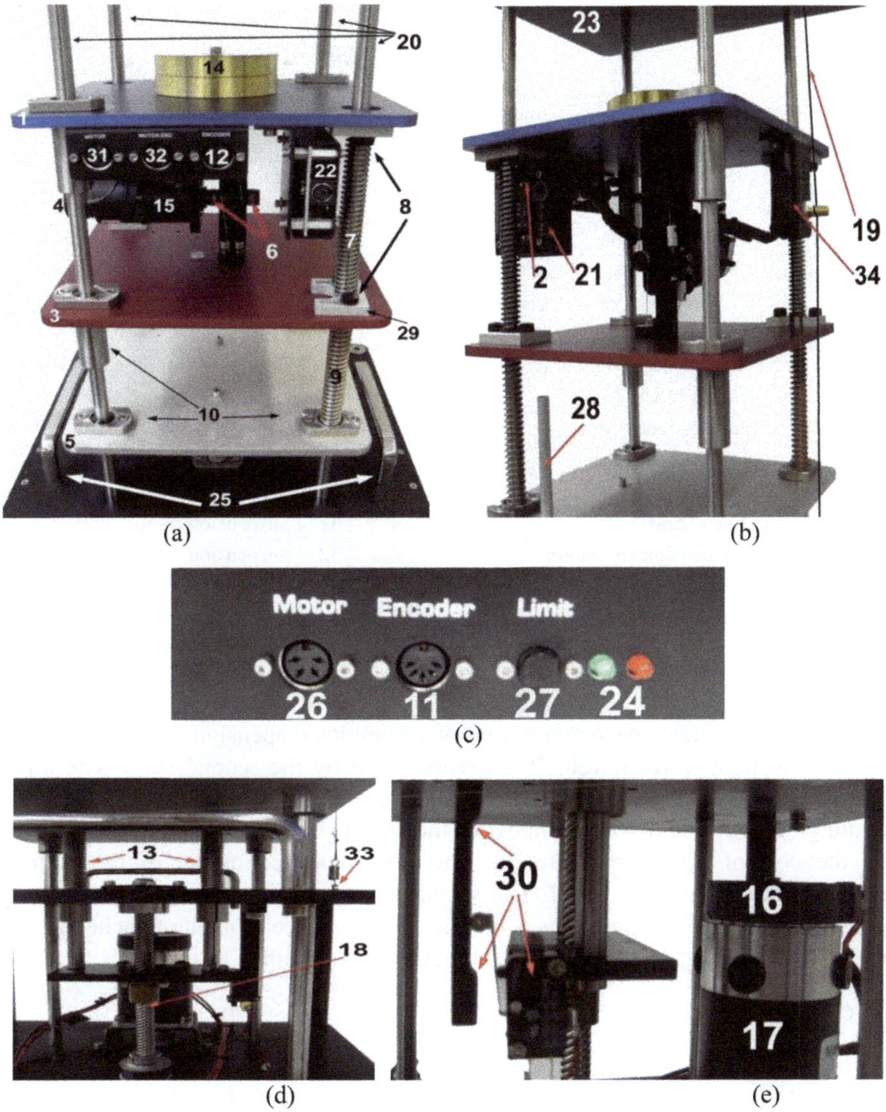

Fig. 4.9 Quanser suspension plant: (**a**) *front top panel view*, (**b**) Quanser suspension system *side view*, (**c**) Quanser suspension plant. *Front bottom panel view*, (**d**) Quanser suspension system *bottom view*, (**e**) Quanser suspension system *bottom view*

In this model, four energy storage elements are present. So the four state variables, inputs and the outputs of the system can be written as follows:

$$x = \begin{vmatrix} z_2 - z_1 \\ \dot{z}_2 \\ z_2 - q \\ \dot{z}_1 \end{vmatrix}, u = \begin{vmatrix} \dot{q} \\ f_c \end{vmatrix}, \quad y = \begin{vmatrix} z_2 - z_1 \\ \ddot{z}_2 \end{vmatrix} \tag{4.18}$$

Table 4.3 Nomenclature of Quanser suspension system component

ID #	Description	ID #	Description
1	Top plate (blue, vehicle body mass)	18	Lead screw
2	Accelerometer gain potentiometer	19	Encoder thread
3	Middle plate (red vehicle tyre mass)	20	Stainless steel shafts
4	Suspension encoder	21	Accelerometer
5	Bottom plate (white, road)	22	Accelerometer connector
6	Suspension motor capstan able	23	Plant top cover
7	2 Adjustable springs (vehicle suspension springs)	24	Limit switch safety lights
8	Spring holder set screw	25	Plant handles
9	2 Adjustable springs (vehicle tyre springs)	26	Bottom plate motor connector
10	Linear bearing blocks	27	Limit switch push key
11	Bottom plate encoder connector	28	Safety rod
12	Top plate encoder connector	29	Movable spring holders
13	Bottom plate counter weight springs	30	Safety limit switch
14	Payload mass (brass)	31	Suspension motor connector
15	Quanser suspension DC motor	32	Suspension encoder connector
16	Bottom plate encoder	33	Encoder thread anchor
17	Bottom plate servo motor	34	Top plate encoder

where the first state-space representation resembles suspension travel/deflection. The vehicle body vertical velocity is represented by the second state. The third state-space equation stands for the tyre deflection which is a measure of road handling. The first and second inputs to the system are the road surface velocity and the control action, respectively. The first measured output of the system represents the suspension travel. The second measured output of the system will be the body acceleration, and this is measured by the accelerometer attached to the vehicle body. The state-space representation of the system is as follows:

$$A = \begin{vmatrix} 0 & 1 & 0 & -1 \\ -\frac{k_2}{m_2} & -\frac{c_0}{m_2} & 0 & \frac{c_0}{m_2} \\ 0 & 0 & 0 & 1 \\ \frac{k_2}{m_1} & \frac{c_0}{m_1} & \frac{k_1}{m_1} & -\frac{c_0 + c_t}{m_1} \end{vmatrix}, \quad B = \begin{vmatrix} 0 & 0 \\ 0 & \frac{1}{m_2} \\ -1 & 0 \\ \frac{c_t}{m_1} & -\frac{1}{m_1} \end{vmatrix}, \tag{4.19}$$

$$C = \begin{vmatrix} 1 & 0 & 0 & 0 \\ -\frac{k_2}{m_2} & -\frac{c_0}{m_2} & 0 & \frac{c_0}{m_2} \end{vmatrix}, \quad D = \begin{vmatrix} \frac{0}{0} & 0 \\ 0 & \frac{1}{m_2} \end{vmatrix}$$

4.4.1.2 Experimental Setup

In this chapter, the Quanser suspension system has been set up to investigate various skyhook control strategies. Figures 4.10 and 4.11 show the experimental setup of the Quanser suspension system and the MATLAB/Simulink model of the Quanser plant, respectively.

Table 4.4, below, lists and characterizes the main parameters related to the Quanser suspension system. Some of these parameters have been used to obtain the system equations of motion (EOM) and the mathematical modelling of the plant.

The FAULHABER DC-micro motor series 3863H048C 1721 has been used as the actuator of the Quanser suspension plant. The actuator motor specification is given in Table 4.5. Figure 4.12 depicts the DC-micro motor characteristics curve. For a specific torque, the speed, current, output power and efficiency of the DC motor can be read from this figure.

Fig. 4.10 The Quanser quarter-car model experimental setup

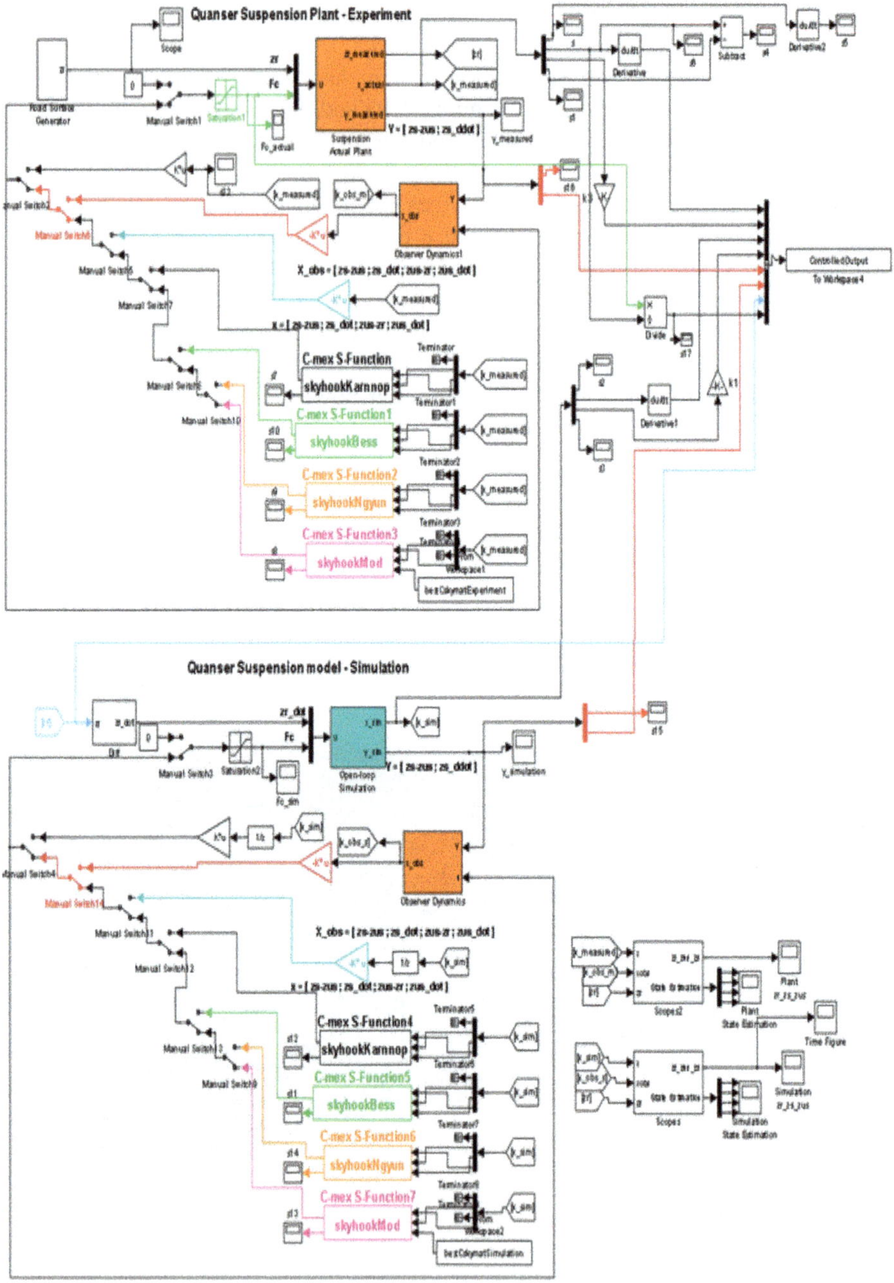

Fig. 4.11 The Quanser suspension plant modelled in Simulink

Table 4.4 Nominal parameter values used in the experiment

Parameter	Description	Value	Unit
m_1	Total mass of middle plate (red) with attached equipment	1	kg
m_2	Top plate (blue) with attached equipment	2.5	kg
k_1	Tyre spring coefficient	2500	N/m
k_2	Suspension spring coefficient	900	N/m
c_0	Suspension damping coefficient	7.5	N-s/m
c_t	Tyre damping coefficient	5	N-s/m
	Suspension motor torque constant	0.115	Nm/A
	Suspension motor shaft radius	0.006	m
	Suspension encoder resolution	0.00094	m/count
	Bottom (white) plate encoder resolution	0.00022	m/count
	Middle (red) plate encoder resolution	0.00049	m/count
	Suspension travel range	0.038	m
	Middle (red) plate travel range	0.03	m
	Bottom (white) plate travel range	0.036	m
	Accelerometer sensitivity	9.81	$m/s^2/v$

The reaction of the road profile input in the simulation and experiment are shown for the passive and four semi-active suspension systems described above is depicted in Fig. 4.13. Random 5 s along the total length of the simulation is shown for better visualization.

The comparison of the results is discussed in the next subsection.

4.4.1.3 Comparison of Ride Comfort Performance

The quantitative evaluation of the ride comfort performance (RCP) has been carried out following the methodology described in Sect. 4.4.1.1. The NRMS achieved in the simulation and experiment by the passive and four semi-active suspension systems are shown in Fig. 4.15.

Figure 4.14a, b shows the comparison of NRMS of the control systems in the simulation environment and experimental setup, respectively. The improvements on comfort index NRMS of the proposed modified skyhook control, optimal skyhook control of Nguyen et al. [51], modified skyhook control of Bessinger et al. [15] and continuous skyhook control of Karnopp et al. [14] with respect to the passive system are 20.34%, 9.77%, 7.64% and 9.63% in the simulation environment and 19.06%, 10.54%, 8.03% and 9.62% in the experimental setup, respectively.

4.4.1.4 Comparison on Road-Handling Performance

The evaluation of the road-handling performance has been done according to Sect. 4.3.1.2. Figure 4.6 shows the comparison of the RHP of the semi-active control algorithms.

Table 4.5 The FAULHABER DC-micro motor specification [144]

Parameter	Description	Value	Unit
U_N	Nominal voltage	48	V
R	Terminal resistance	2.58	Ω
$P_{2\ max.}$	Output power	217	W
$\eta_{max.}$	Efficiency, max.	86	%
n_0	No-load speed	5800	rpm
I_0	No-load current (with shaft ø 6 mm)	0.084	A
M_H	Stall torque	1461	mNm
M_R	Friction torque	6.5	mNm
k_n	Speed constant	120	rpm/V
k_E	Back-EMF constant	8.33	mV/rpm
k_M	Torque constant	79.7	mNm/A
k_I	Current constant	0.013	A/mNm
$\Delta n/\Delta M$	Slope of *n–M* curve	3.9	rpm/mNm
L	Rotor inductance	700	μH
τ_m	Mechanical time constant	4.7	ms
J	Rotor inertia	115	gcm^2
$\alpha_{max.}$	Angular acceleration	127	$\cdot 10^3$ rad/s^2
$R_{th\ 1}$ / $R_{th\ 2}$	Thermal resistance	2.5/6	K/W
τ_{w1} / τ_{w2}	Thermal time constant	50/900	s
	Operating temperature range		
	– Motor	−30 ... +125	°C
	– Rotor, max. permissible	+155	°C
	Shaft bearings	Ball bearings	
	Shaft load max.		
	– With shaft diameter	6	mm
	– Radial at 3000 rpm (3 mm from bearing)	60	N
	– Axial at 3000 rpm	6	N
	– Axial at standstill	50	N
	Shaft play		
≤	– Radial	0.015	mm
=	– Axial	0	mm
	Housing material	Steel black&$$$;	
	Weight	390	g
	Direction of rotation	Clockwise	

Figures 4.15a and b show the comparison of RHP of the control systems. The RHP of the proposed modified skyhook control and the optimal skyhook control of Nguyen et al. increases by 8.52%, 16.15% in a simulation environment and 10.07%, 16.75% in the experimental setup, respectively, compared to the passive suspension system, whereas RHPs of the modified skyhook control of Bessinger et al. [15] and the continuous skyhook control of Karnopp et al. [14] decrease by 6.38%, 1.34% in

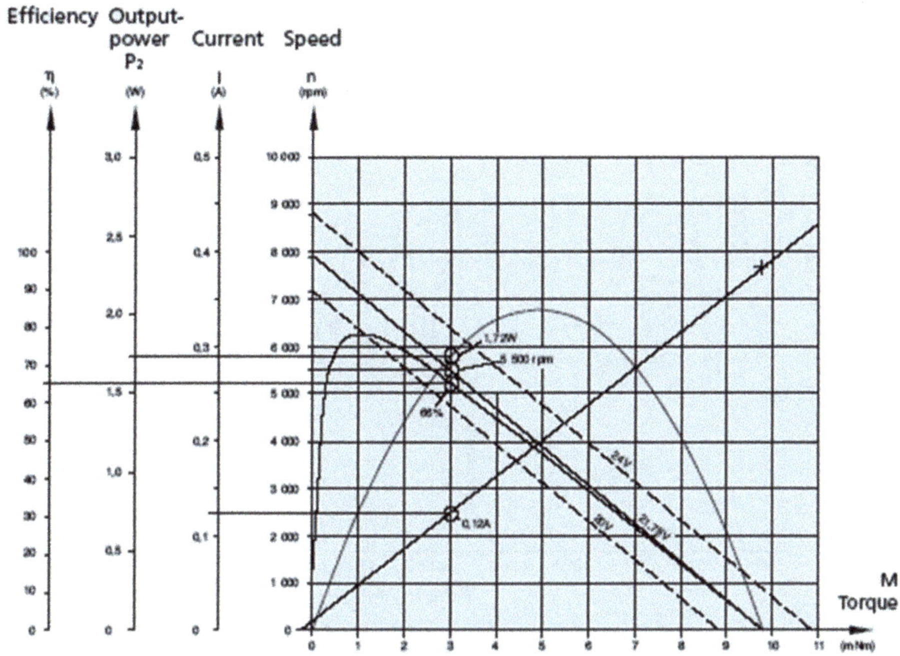

Fig. 4.12 DC-micro motor characteristics curve [144]

the simulation environment and 6.37%, 1.05% in the experimental setup, respectively, compared to the passive suspension system.

4.4.1.5 Human Vibration Sensitivity

To perform the human vibration sensitivity test, the vertical vibration of car suspension in frequency domain has been evaluated and shown in Fig. 4.16.

4.4.1.6 Admissible Acceleration Levels Test Based on ISO 2631

The admissible acceleration levels test has been done according to Sect. 4.3.2.2. Here the Quanser quarter-car model has been used to determine the WRMS value of the sprung-mass acceleration. The value of WRMS is 0.76 in experimental analysis and 0.67 in the simulation environment on road class C. This confirms the validation of the proposed modified skyhook control system in terms of ride comfort enhancement for two degrees of freedom semi-active suspension system.

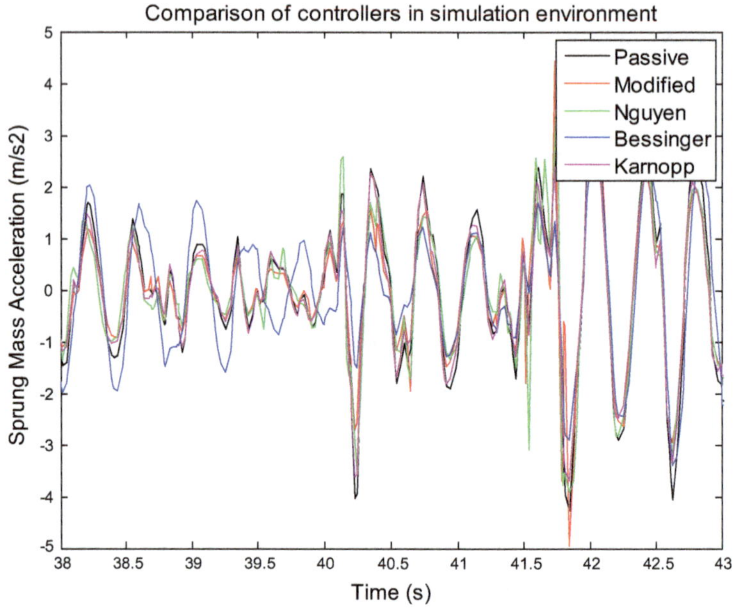

(a)

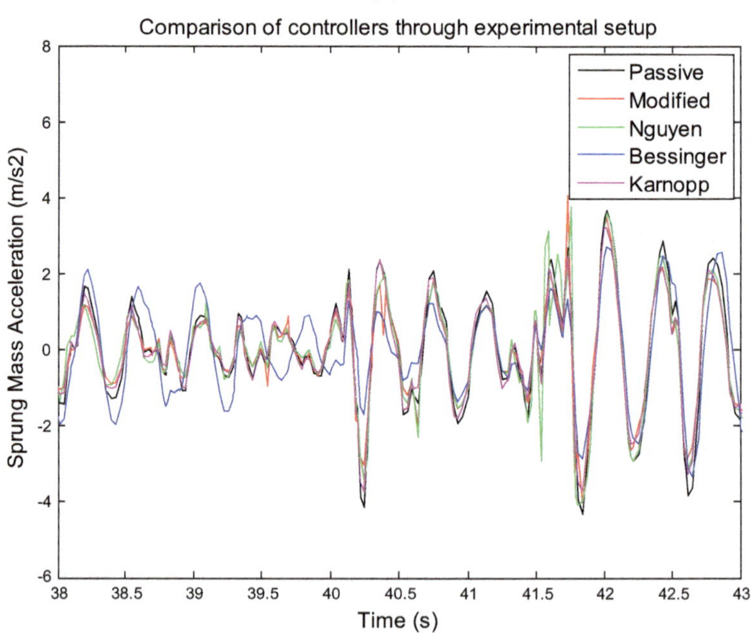

(b)

Fig. 4.13 The sprung-mass acceleration of the passive and semi-active suspension systems (**a**) in a simulation environment, (**b**) in the experimental setup

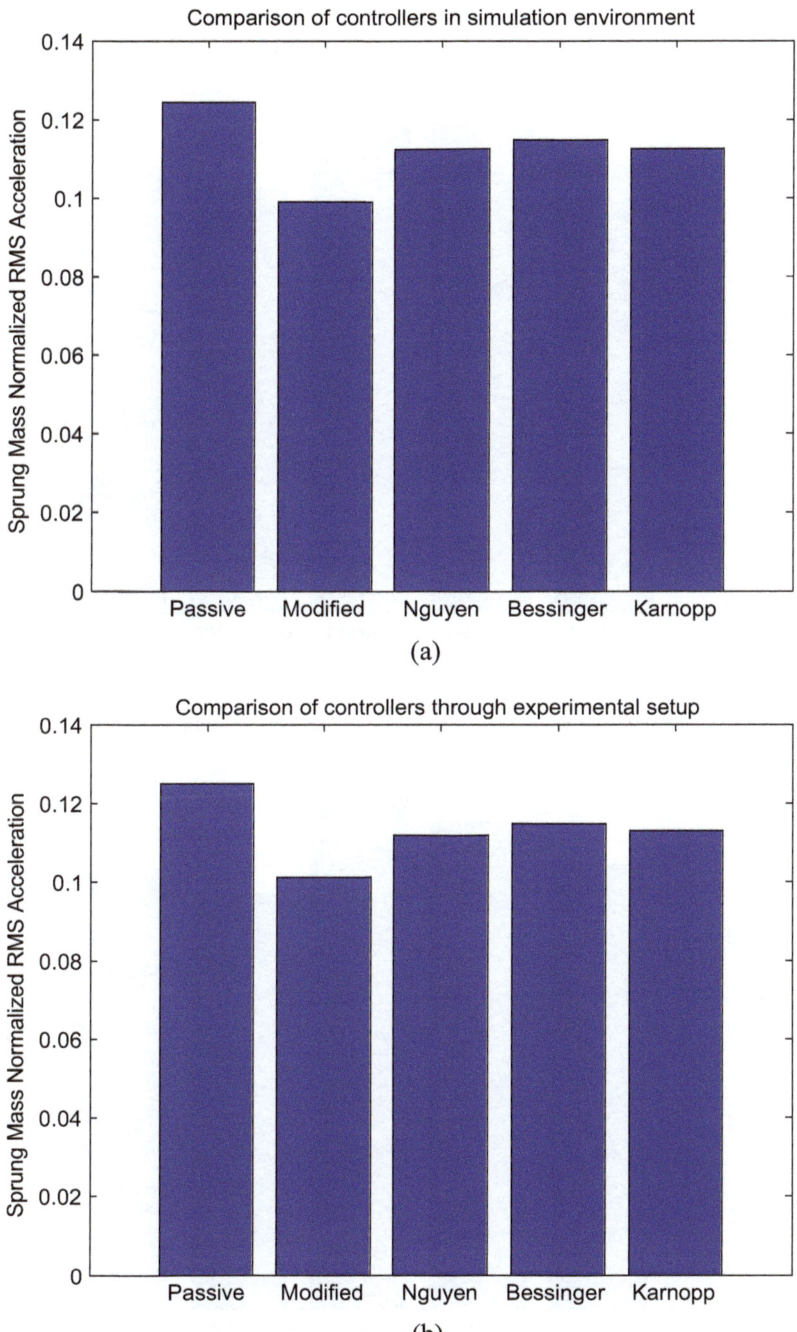

Fig. 4.14 The ride comfort performance comparison (**a**) in simulation environment, (**b**) through experimental setup

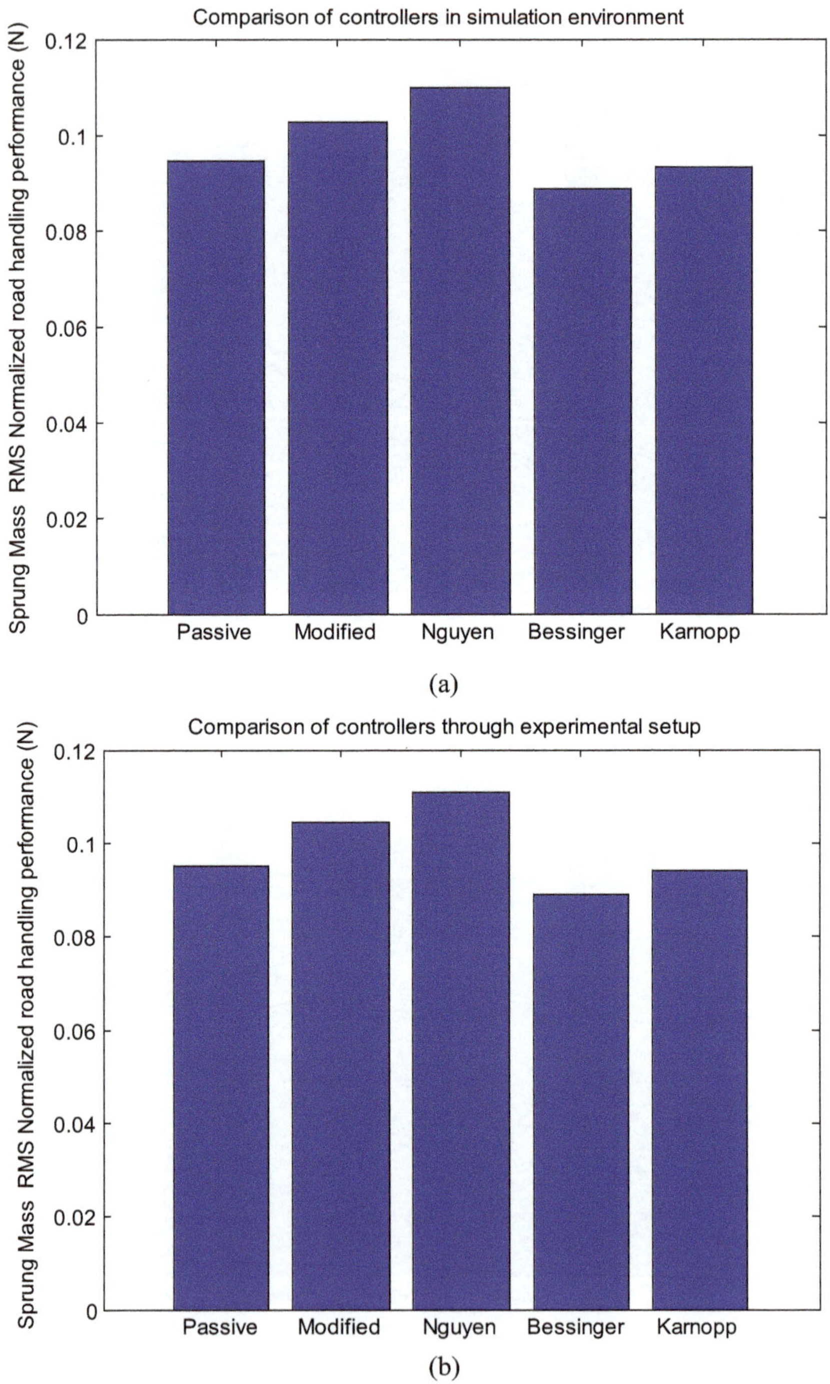

Fig. 4.15 The road-handling performance comparison (**a**) in simulation environment, (**b**) through experimental setup

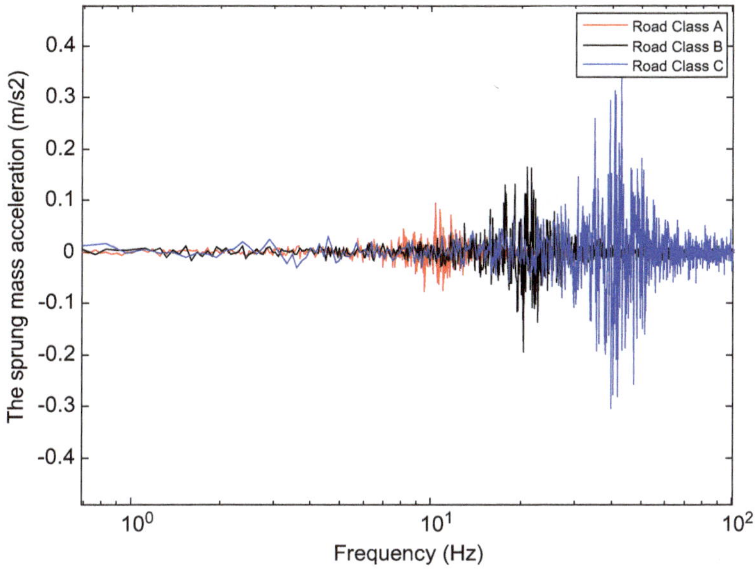

Fig. 4.16 Vertical vibration of car suspension in frequency domain

4.5 Conclusions

In this chapter, a brief discussion on the proposed modified skyhook control approach, optimal skyhook control of Nguyen et al. [51], modified skyhook control of Bessinger et al. [15] and continuous skyhook control of Karnopp et al. [14] has been presented. A road profile has been generated to study the performance of the different controllers. The two degrees of freedom quarter-car model described in Chap. 3 has been simulated to compare the controllers' performances. The Quanser quarter-car suspension plant has been also used to compare the performance of the controllers in an experimental environment. From the above simulation and experimental analysis, it can be concluded that the proposed modified skyhook control strategy provides the best performance of those investigations because it decreases sprung-mass acceleration to a great extent compared to the passive system and other skyhook controllers described in the literature. This system also keeps the road-handling performance of the vehicle within a range which is acceptable by ISO 2631 standards. These models have also been evaluated in terms of human vibration perception and admissible acceleration levels based on ISO 2631 in this chapter.

Chapter 5
Full Car Model Cornering Performance

Abstract In this chapter, a dynamic model of a full car which considers the road bank angle is developed. The first section describes the full car model design along with the vehicle tilting model. The vehicle rollover estimation procedure is described in Sect. 5.1. Section 5.2 describes the controller design that is required to control the vehicle tilt while cornering. The next section is comprised of descriptions of the road profiles and driving scenarios that will be used in simulation and experimental analysis in the next two chapters. The evaluation criteria are described in the last section to compare the results of different controllers in terms of ride comfort, admissible acceleration level test based on ISO 2631 and road handling performance.

5.1 Full Car Modelling

The car suspension behaviour can be expressed in many ways; the full car model is one of them. The full vehicle's mathematical model is developed in this section. The semi-active suspension system has been employed in this model. The vehicle model designed in this manuscript has nine degrees of freedom, and those are the heave modes of four wheels and the lateral, roll, heave, pitch and yaw modes of the vehicle body.

5.1.1 Semi-active Suspension Model

A schematic diagram is shown in Fig. 5.1 which emulates a full vehicle semi-active suspension system.

The model has four unsprung masses $m_1 \sim m_4$ (front-left, front-right, rear-left and rear-right wheels) connected to the car body or a single sprung mass m. Each quarter of the suspension system has a damping valve, an actuator and a spring which are connected in parallel. The spring and the damper are employed to suppress high-frequency vibrations above the bandwidth of the force generator. The tyres are modelled without any damping components. It is modelled as a simple linear. In this model m represents the sprung mass, k_{i1} (i = 1,2,3,4) is the tyre stiffness coefficient or tyre spring constant of each tyre and k_{i2} (i = 1,2,3,4) is the suspension stiffness or suspension spring constant. f_{di} (i = 1,2,3,4) is the actuating

S. Kashem et al., *Vehicle Suspension Systems and Electromagnetic Dampers*,
Springer Tracts in Mechanical Engineering, DOI 10.1007/978-981-10-5478-5_5

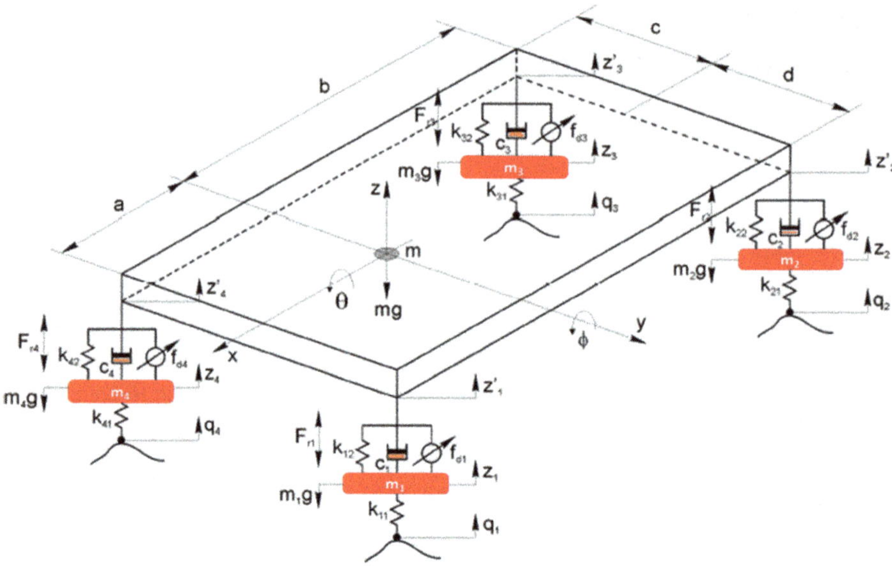

Fig. 5.1 A schematic diagram of a full vehicle active suspension system [142]

force. c_1, c_2, c_3, c_4 are the suspension damping constants; q_i, z_i, z'_i $(i = 1{,}2{,}3{,}4)$ represents road profile input, displacement of unsprung mass and displacement of sprung mass, respectively.

The global coordinate system of this model has been defined in such a way that the x and y axes are aligned with the vehicle longitudinal and lateral motion direction, respectively. The z axis is defined to be normal to the road surface. The x, y and z axes are fixed at the vehicle's centre of gravity (in the plane shown in Fig. 5.1). The pitch angle and roll angle are represented by ϕ and θ, respectively.

In this manuscript, a ride model was derived based on the work done by Ikenaga et al. [145, 146], Wang et al. [33] and Bakar et al. [97]. The equations of motion for this system are:

$$\left\{\begin{array}{l} m_1\ddot{z}_1 = k_{11}(q_1 - z_1) + k_{12}\left(z'_1 - z_1\right) + c_1\left(\dot{z}'_1 - \dot{z}_1\right) + f_{\mathrm{d}1}; \\ m_2\ddot{z}_2 = k_{21}(q_2 - z_2) + k_{22}\left(z'_2 - z_2\right) + c_2\left(\dot{z}'_2 - \dot{z}_2\right) + f_{\mathrm{d}2}; \\ m_3\ddot{z}_3 = k_{31}(q_3 - z_3) + k_{32}\left(z'_3 - z_3\right) + c_3\left(\dot{z}'_3 - \dot{z}\right)_3 + f_{\mathrm{d}3}; \\ m_4\ddot{z}_4 = k_{41}(q_4 - z_4) + k_{42}\left(z'_4 - z_4\right) + c_4\left(\dot{z}'_4 - \dot{z}_4\right) + f_{\mathrm{d}4}; \\ m\ddot{z} = k_{12}\left(z_1 - z'_1\right) + k_{22}\left(z_2 - z'_2\right) + k_{32}\left(z_3 - z'_3\right) + k_{42}\left(z_4 - z'_4\right) + c_1\left(\dot{z}_1 - \dot{z}'_1\right) \\ +c_2\left(\dot{z}_2 - \dot{z}'_2\right) + c_3\left(\dot{z}_3 - \dot{z}'_3\right) + c_4\left(\dot{z}_4 - \dot{z}'_4\right) - f_{\mathrm{d}1} - f_{\mathrm{d}2} - f_{\mathrm{d}3} - d_{\mathrm{d}4}; \\ J_X\ddot{\theta} = -\left[k_{32}\left(z_3 - z'_3\right) + c_3\left(\dot{z}_3 - \dot{z}'_3\right) + k_{42}\left(z_4 - z'_4\right) + c_4\left(\dot{z}_4 - \dot{z}'_4\right)\right]c+ \\ +\left[k_{12}\left(z_1 - z'_1\right) + c_1\left(\dot{z}_1 - \dot{z}'_1\right) + k_{22}\left(z_2 - z'_2\right) + c_2\left(\dot{z}_2 - \dot{z}'_2\right)\right]d+ \\ +(f_{\mathrm{d}1} + f_{\mathrm{d}2})d - (f_{\mathrm{d}3} + f_{\mathrm{d}4})c + M_{\mathrm{r}}; \\ J_Y\ddot{\phi} = -\left[k_{12}\left(z_1 - z'_1\right) + c_1\left(\dot{z}_1 - \dot{z}'_1\right) + k_{42}\left(z_4 - z'_4\right) + c_4\left(\dot{z}_4 - \dot{z}'_4\right)\right]a+ \\ +\left[k_{22}\left(z_2 - z'_2\right) + c_2\left(\dot{z}_2 - \dot{z}'_2\right) + k_{32}\left(z_3 - z'_3\right) + c_3\left(\dot{z}_3 - \dot{z}'_3\right)\right]c+ \\ +(f_{\mathrm{d}2} + f_{\mathrm{d}3})b - (f_{\mathrm{d}1} + f_{\mathrm{d}4})a; \end{array}\right. \tag{5.1}$$

where

$$\begin{aligned} z_1' &= z - (a\phi - d\theta) \\ z_2' &= z + (b\phi + d\phi) \\ z_3' &= z + (b\phi - c\phi) \\ z_4' &= z - (a\phi + c\phi) \end{aligned} \tag{5.2}$$

Here I_x, I_y and I_z are the tilt moment of inertia, pitch moment of inertia and yaw moment of inertia of the vehicle, respectively. M_r is the manoeuvring torque to roll, and it has been introduced as a disturbance signal of the model in Eq. (5.1). It has been assumed that the roll moment M_{steer} caused by the steering manoeuvre is the only source of vehicle disturbance [33], i.e. $M_r = M_{steer}$, where

$$M_{steer} = m_t\left(\ddot{y} + V\dot{\psi}\right)h_t. \tag{5.3}$$

Here m_t represents the total mass of the vehicle, y is the lateral displacement of vehicle body, V is the longitudinal velocity of the vehicle, $\dot{\psi}$ represents the yaw rate and h_t is the distance between the roll centre and the centre of gravity of the vehicle.

5.1.2 Vehicle Tilting Model

A model describing the relationship between drivers' steering commands and intended tilting angle is derived in this manuscript. A vehicle tilting model has been derived and is based on the work done by Piyabongkarn et al. [32], Kidane et al. [31], Rajamani [109] and Sang-Gyun et al. [39, 147]. Figure 5.2 indicates a free body diagram of a narrow two-wheeled vehicle model (such as a tilting bicycle model) which is travelling around a banked corner. In this model, the global coordinate system is characterized in such a way that the y axis is aligned with the road surface. The z axis is characterized to be normal to the road surface. In the plane shown in Fig. 5.2, the z and y axes are fixed at the vehicle's centre of gravity.

Here β is the road bank angle, $\dot{\psi}$ refers to the yaw rate, $\ddot{y}$ refers to the lateral position acceleration and the longitudinal vehicle velocity is represented by V.

The nonlinear dynamic model of a vehicle tilting model which considers the road bank angle can be represented by the following equation.

$$m\ddot{y} + m\dot{\psi}V + \ddot{\theta}mh\cos(\theta) - m\dot{\theta}^2 h\sin(\theta) = F_f + F_r + mg\sin(\beta)$$

$$I_z\ddot{\psi} = l_f F_f - l_r F_r$$

$$I_x\ddot{\theta} = \left(-\ddot{\theta}hm\sin(\theta) - \dot{\theta}\,hm\cos(\theta) + mg\cos(\beta)\right)h\sin(\theta) - (F_f + F_r)h\cos(\theta) + T \tag{5.4}$$

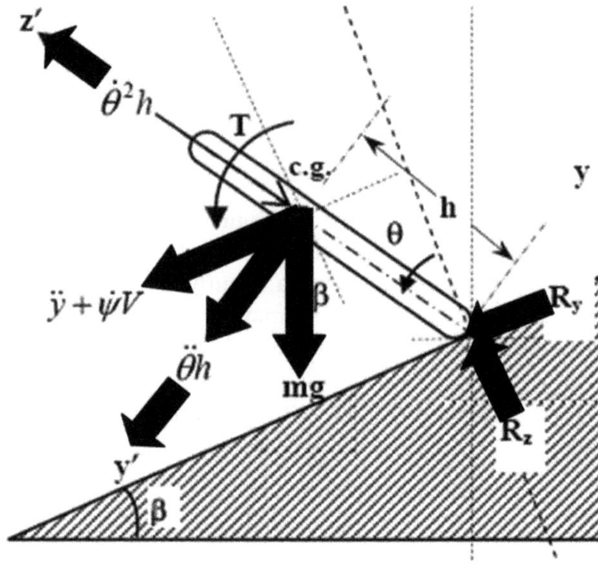

Fig. 5.2 Free body diagram of a bicycle model [31]

The nonlinear model stated above can be represented by the following linear model using small angle approximations.

$$m\ddot{y} + m\dot{\psi}V + \ddot{\theta}mh = F_{\mathrm{f}} + F_{\mathrm{r}} + mg\beta \tag{5.5}$$

$$I_z\ddot{\psi} = l_{\mathrm{f}}F_{\mathrm{f}} - l_{\mathrm{r}}F_{\mathrm{r}} \tag{5.6}$$

$$I_x\ddot{\theta} = mgh\theta - (F_{\mathrm{r}} + F_{\mathrm{f}})h + T \tag{5.7}$$

The lateral tyre force is proportional to wheel camber angle and the wheel slip angle. Since all the wheels don't tilt with the vehicle model during turns (only the sprung mass tilt), the camber angle of the wheels is equal to zero. Thus the front and rear wheel lateral forces can be represented by the following Eq. (5.8).

$$\begin{aligned} F_{\mathrm{f}} &= 2C_{\mathrm{f}}\alpha_{\mathrm{f}} = 2C_{\mathrm{f}}\left(\delta - \frac{\dot{y} + l_{\mathrm{f}}\dot{\psi}}{V}\right). \\ F_{\mathrm{r}} &= 2C_{\mathrm{r}}\alpha_{\mathrm{r}} = 2C_{\mathrm{r}}\left(-\frac{\dot{y} - l_{\mathrm{r}}\dot{\psi}}{V}\right). \end{aligned} \tag{5.8}$$

Here δ is the steering input to front wheels; C_{f} and C_{r} are the front and rear wheels cornering stiffness, respectively. In the above equation, the tyre forces are multiplied by a factor two because the tilting vehicle considered in this manuscript has two front wheels and two rear wheels (it is a full car model not a bicycle model).

5.2 Vehicle Rollover Estimation

This section describes the basic dynamics of a vehicle tip-over. The objective is to present a justification of the direct tilt control manoeuvre used to minimize lateral acceleration of the vehicle, which will be developed in the next section. Vehicle tip-over indicates that the vehicle's dynamic stability has become compromised, signifying a possible rollover if the unbalanced forces on the vehicle continue to increase. It would be feasible to effectively measure and assess the vehicle's stability, if the dynamics leading up to rollover is well understood. First of all, it is necessary to explain vehicle tip-over mechanics under simulated static load conditions.

Static stability is a measure of a vehicle's tendency to tip-over under simulated lateral acceleration conditions. A basic calculation can be done where the fundamental forces and moments leading to vehicle rollover are summed to roughly estimate the vehicle's rollover threshold. In Fig. 5.3, Allen et al. [148] illustrated these basic forces. If the vehicle is level with the ground, its rollover threshold becomes a function of the vehicle's track width (t) and centre of gravity height (h) as displayed in the following equation:

$$\frac{a_y}{g} = \frac{t}{2h} \tag{5.9}$$

where

a_y = Lateral acceleration
g = Gravitational acceleration
t = Vehicle track width
h = Height of the vehicle's CG above the ground

The rollover threshold defines the maximum lateral acceleration value over gravity (a_y/g) that the vehicle would be able to reach before tipping over. To improve the rollover threshold of the vehicle, the vehicle's track width can be increased, or the centre of gravity height can be reduced. The rollover threshold is also known as the static stability factor (SSF). The static stability analysis is needed to generate a conservative rollover threshold estimate for the vehicle.

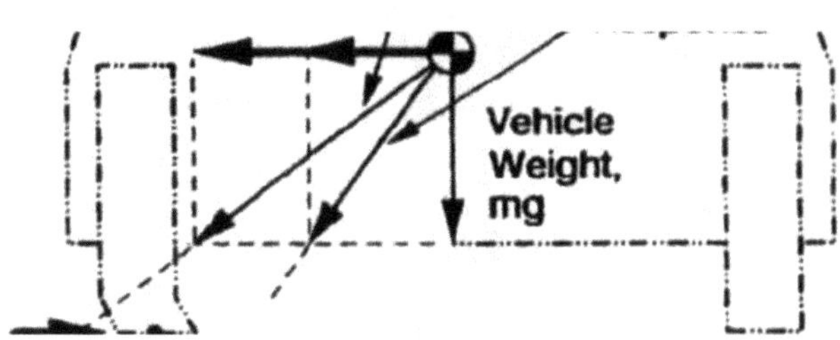

Fig. 5.3 Stable and unstable lateral forces acting on a static vehicle [148]

5.3 Controller Design

The proposed modified skyhook control strategy described in Chap. 4 has been used in designing the semi-active suspension system of the full car model. The velocity of the sprung mass and the unsprung mass of each wheel of the full car model is used as the input to the controller. The required force has been calculated for each wheel individually for better ride comfort and ride handling. Desired tilt angle and corresponding actuator force to tilt the vehicle have been calculated separately with the direct tilt control strategy. This strategy has been described in the next section.

5.3.1 Direct Tilt Control Design

The driver of the vehicle has full control of the front wheel steering angle in the direct tilt control system. Hence the vehicle's tilt control actuator requires to tilt the vehicle to the desired angle according to the trajectory the vehicle's driver is driving. The system also needs to maintain the vehicle's stability.

5.3.1.1 Desired Tilting Angle

In this manuscript, the "desired tilting angle" (θ_{des}) is referred to as the unique tilting angle for the vehicle that is able to keep the vehicle stable for a given situation of road bank angle, speed, pitch, roll and yaw rate. The tilt stability of the vehicle is ensured when the summation of all the steady-state moments is zero which acts at the vehicle's centre of gravity. Therefore replacing the summation of all the lateral forces and equating the right side of Eq. (5.7) to zero from Eq. (5.5) give Eq. (5.10) as follows:

$$mgh\theta - \left(m\ddot{y} + m\dot{\psi}V + \ddot{\theta}mh - mg\beta\right)h + T = 0 \tag{5.10}$$

To find the solution of the above second-order differential equation, the desired tilt angle of the vehicle needs to be determined. The vehicle's angular acceleration and the torque of the actuator have been set to zero, and the tilt angle (roll angle) θ is derived. The tilt angle achieved in Eq. (5.11) is the target vehicle's desired tilt angle.

$$\theta_{\text{des}} = \underbrace{\frac{\ddot{y} + \dot{\psi}V}{g}}_{a} - \underbrace{\beta}_{b} \tag{5.11}$$

To determine the vehicle's desired tilt angle, the road bank angle plays an important role as shown in Eq. (5.11). It also indicates that a non-zero tilt angle is needed to keep the vehicle stable while the vehicle is travelling on a straight road (steering angle input is zero) having a certain bank angle even though the vehicle's lateral position accelerates and the yaw rate remains zero.

5.3.1.2 Desired Actuator Force

According to Newton's second law for rotation, torque is equal to a moment of inertia time's angular acceleration,

$$T = I_x * \ddot{\theta}_{\mathrm{f}} \tag{5.12}$$

where I_x is the tilt or roll moment of inertia of the vehicle and $\theta_{\mathrm{f}} = \theta_{\mathrm{des}}$ is the desired tilt or roll angle of the vehicle. Again in physics a force multiplied by a moment arm is equal to torque,

$$T = f_{\mathrm{dtotal}} * w \tag{5.13}$$

where f_{dtotal} is the total required force to tilt the vehicle and w is the half of the wheel base. In the full car model, distances from CG to the right wheel and left wheel are c and d, respectively, since the two actuators of the same side of the vehicle have to produce the torque, according to Fig. 5.4b. Thus Eq. (5.13) yields

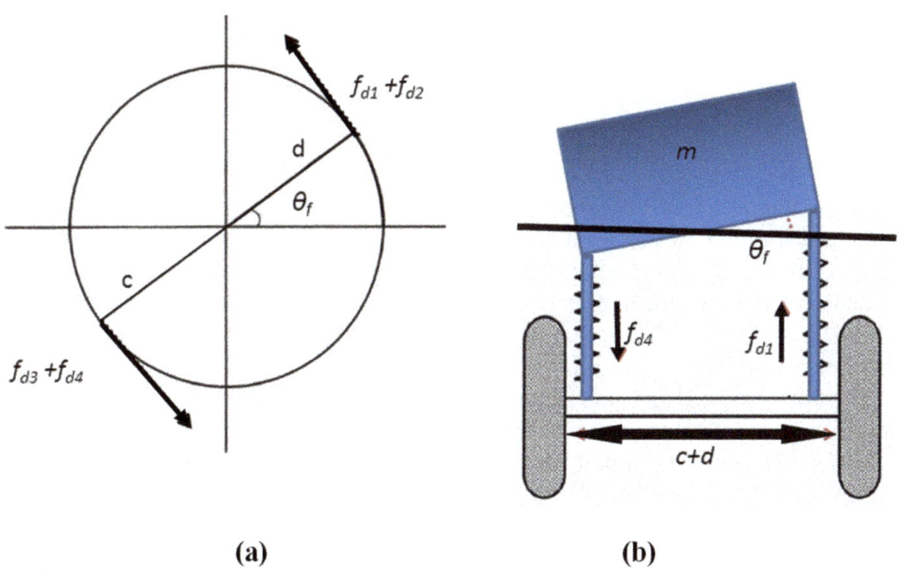

Fig. 5.4 (**a**) Acting torque on the vehicle body, (**b**) *front view* of the tilting vehicle

$$T = (f_{d1} + f_{d2}) * d + (f_{d3} + f_{d4}) * c \tag{5.14}$$

Here distances from CG to the right wheel and left wheel are equal ($c = d$), and all the actuators have to produce equal values of force. So inserting $f_{dt} = f_{d1} = f_{d2} = f_{d3} = f_{d4}$ and $d = c$ in Eq. (5.14) yields

$$T = 4f_{dt} * c \tag{5.15}$$

Here f_{dt} is the required force by a single actuator to tilt the vehicle. Thus Eqs. (5.12) and (5.15) yield

$$f_{dt} = \frac{I_x * \ddot{\theta}_f}{4c} \tag{5.16}$$

5.3.1.3 Actuator Selection

The challenge of selecting the actuator with the required force to tilt the vehicle has been resolved by the algorithm given below:

$$\begin{aligned} f_{d1} = f_{d2} &= \begin{cases} f_{dt} & \text{if } \theta_{des} > 0 \,\&\, \frac{a_y}{g} < \frac{t}{2h} \\ 0 & \text{if } \theta_{des} \le 0 \text{ or } \frac{a_y}{g} > \frac{t}{2h} \end{cases} \\ f_{d3} = f_{d4} &= \begin{cases} 0 & \text{if } \theta_{des} \ge 0 \text{ or } \frac{a_y}{g} > \frac{t}{2h} \\ f_{dt} & \text{if } \theta_{des} < 0 \,\&\, \frac{a_y}{g} < \frac{t}{2h} \end{cases} \end{aligned} \tag{5.17}$$

According to the full car model described in Sect. 5.1.1, $\theta_{des} > 0$ indicates that vehicle is turning to the right. Thus the vehicle is required to tilt towards the positive side of the roll angle. Hence, the required force should be given to the actuators which are located at the opposite side of the turning centre. Similarly when θ_{des} is <0, it indicates that the vehicle is turning to the left, and the right side actuator is required to generate *a* force equal to f_{dt}.

For both cases, if the maximum lateral acceleration value over gravity (a_y/g) becomes greater than the rollover threshold ($t/2h$) as described in Sect. 5.2, then no extra force would be exerted on the actuator via direct tilt control to avoid the vehicle rollover. When the proposed skyhook control and direct tilt control both are activated, f_{dt} is simply added to the skyhook damping force.

5.4 Road Profile and Driving Scenario

In order to evaluate the performance of the full car model with the proposed control algorithms, the simulation of the model needs to be run on road surfaces that are defined by ISO to test a vehicle performance. In this manuscript, the full car model experienced road classes A, B and C defined by ISO8608 (described in Sect. 4.2).

To investigate how the controller behaves under different road conditions, a road profile is also generated using the three classes (A, B and C) of road profiles where each class lasts for 20 s. Figure 4.3 shows the combined road profile input that has been used in the simulations to compare the performance of the controllers.

To evaluate the performance of the full car model with the proposed DTC control algorithms, the simulation of the model needs to be run on road surfaces with a bank angle. Different steering angle inputs have also been provided to the vehicle model to realize the response of the DTC control. In the next subsections, different typical driving scenarios that have been used in this evaluation are described.

5.4.1 Driving Scenario One

Suppose the vehicle driver is driving the vehicle at a constant speed of 10 m/s. The trajectory of the vehicle is set as a straight line followed by a right turn subject to a steering manoeuvre at $t = 3$ s, and the signal of the steering attains its final value 0.12 radian (11.459°) at $t = 4$ s as shown in Fig. 5.5. The vehicle maintains this steering angle for the rest of the trajectory.

5.4.2 Driving Scenario Two

In this scenario, the steering angle remains zero for the whole trajectory which means there is no turn in the road (Fig. 5.6). But as shown in Fig. 5.7, there is a tilt in

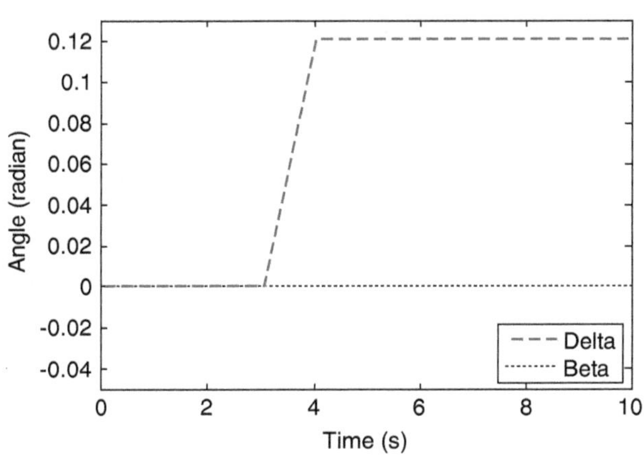

Fig. 5.5 Driving scenario one

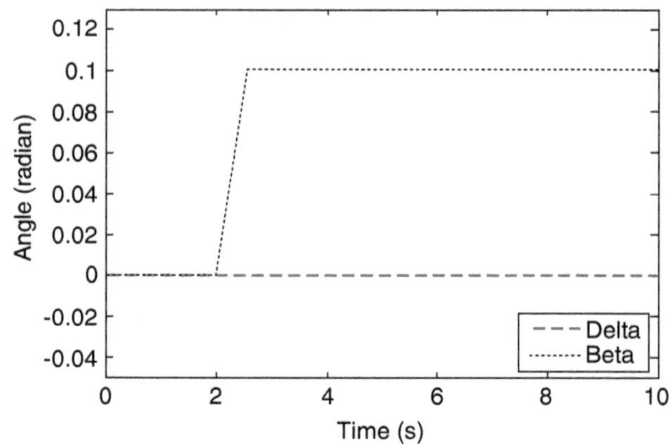

Fig. 5.6 Driving scenario two

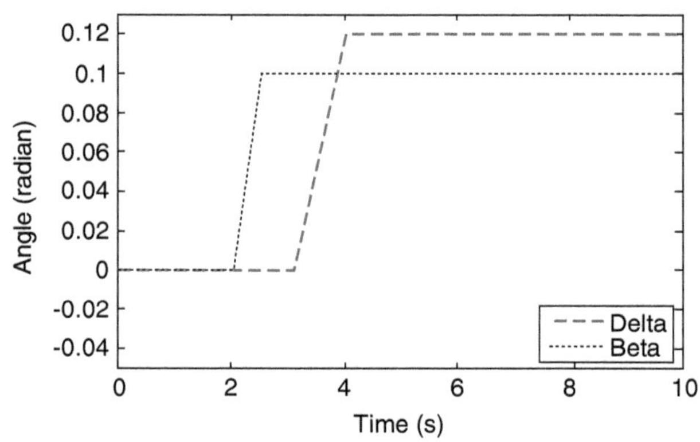

Fig. 5.7 Driving scenario three

the road. The road tilt angle is defined to be zero for the first 2 s and increases gradually to 0.1 radians at $t = 2.5$ s and remains of 0.1 radians for the rest of the vehicle trajectory.

5.4.3 *Driving Scenario Three*

In this scenario, the trajectory of the vehicle is set as a straight line followed by a right turn subject to a steering manoeuvre at $t = 3$ s, and the steering signal attains its final value 0.12 radian (11.459°) at $t = 4$ s as shown in Fig. 5.7. The vehicle maintains this steering angle on the road for the rest of the trajectory. For the first 2 s, the road bank angle is set to zero and increases gradually to 0.1 radians at $t = 2.5$ s and remains 0.1 radian for the rest of the vehicle trajectory.

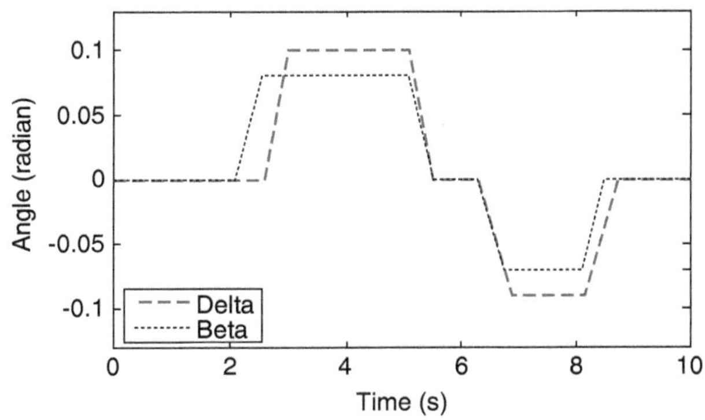

Fig. 5.8 Driving scenario four

5.4.4 Driving Scenario Four

Suppose the vehicle is travelling at the same speed of 10 m/s. The trajectory of the vehicle is set as a straight line followed by a right turn subject to a steering manoeuvre at $t = 2.5$ s, and the steering signal attains its final value 0.1 radian (5.73°) at $t = 3$ s as shown in Fig. 5.8. The vehicle maintains this steering angle for 2 s and starts to run straight from $t = 5.5$ s to $t = 6.25$ s. Then it makes a left turn, and the steering signal attains its final value -0.09 radian at $t = 7$ s. Again it starts to turn the steering angle to 0 radian at $t = 8$ s and follow a straight road for the rest of the trajectory from $t = 8.75$ s.

Road bank is also included at the trajectory to evaluate the tilting performance. As shown in Fig. 5.8, for the first 2 s, the road bank angle is set to zero and increases gradually to 0.8 radians at $t = 2.5$ s and remains at 0.8 radians till $t = 5$ s. The road becomes level at $t = 5.5$ s and starts to bank in the opposite direction at $t = 6.25$ s. The road bank angle attains its final value -0.07 radian at $t = 6.5$ s and remains the same till $t = 8$ s. The road becomes level for the rest of the vehicle trajectory at $t = 8.5$ s. To evaluate the effect of the road bank angle on the tilt control system, the road bank angle is initiated before the introduction of the right turn without the presence of the yaw rate of the vehicle. The skyhook and DTC controlled suspension system are evaluated separately to determine the significance of road bank angle.

5.5 Evaluation Criteria

To evaluate the controllers' performances, ride comfort and road handling performances have been calculated and are presented in the following section.

5.5.1 Evaluation of Ride Comfort Performance

Ride comfort depends on human perceptions of vehicle noise, vibration and motion. It varies from person to person. To determine the ride comfort performance (RCP), the quantitative evaluations of the vehicle body vertical, pitch angular, roll angular and lateral acceleration are carried out following the methodology provided by Abramov et al. [142] and that is the RMS value of the vehicle body vertical acceleration normalized to g (the gravity acceleration). The normalized RMS value of vehicle body vertical acceleration (NV) is calculated using Eq. (5.18).

$$\mathrm{NV} = \left(\frac{1}{T} \int_{t=0}^{T} \left[\frac{\ddot{z}\,(t)}{g} \right]^2 \mathrm{d}t \right)^{1/2} \tag{5.18}$$

Here T is the duration of exposure. Similarly the normalized RMS value of vehicle pitch angular acceleration (NP), roll angular acceleration (NR) and lateral acceleration (NL) is calculated using Eqs. (5.19), (5.20) and (5.21), respectively.

$$\mathrm{NP} = \left(\frac{1}{T} \int_{t=0}^{T} \left[\frac{\ddot{\phi}\,(t)}{g} \right]^2 \mathrm{d}t \right)^{1/2} \tag{5.19}$$

$$\mathrm{NR} = \left(\frac{1}{T} \int_{t=0}^{T} \left[\frac{\ddot{\theta}\,(t)}{g} \right]^2 \mathrm{d}t \right)^{1/2} \tag{5.20}$$

$$\mathrm{NL} = \left(\frac{1}{T} \int_{t=0}^{T} \left[\frac{\ddot{a}_y\,(t)}{g} \right]^2 \mathrm{d}t \right)^{1/2} \tag{5.21}$$

5.5.2 Admissible Acceleration Level Test Based on ISO 2631

The international code ISO 2631 [143] defines a term named weighted RMS acceleration (denoted here as WRMS) which can be formulated by the following equation:

$$\mathrm{WRMS} = \left(\frac{1}{T} \int_{t=0}^{T} \sum_{i=1}^{N} \left[F_i\, \ddot{z}_{2,i}\,(t) \right]^2 \mathrm{d}t \right)^{1/2} \tag{5.22}$$

where $i = 1, 2, 3, N$, T is the duration of exposure and F_i is the weighting coefficients of acceleration in the vertical direction.

5.5.3 Evaluation of Road Handling Performance

To investigate the vehicle's road handling performance, the forces acting on the tyres and the road should be considered. The vehicle's reaction points with the road disturbance are the tyres. For a quarter-car model, the force exchanged between road and tyre is $Fz_1 = k_{11}\,(z_1 - q)$. Then the vehicle's road handling performance (RHP) can be calculated by the root mean square values of these forces normalized to the static forces which act on the wheels. The static forces are represented as $(m_1 + m_2)g$ [142],

$$\text{RHP} = \left(\frac{1}{T}\int_{t=0}^{T}\left[\frac{F_{Z1}(t)}{(m_1 + m_2)g}\right]^2 \mathrm{d}t\right)^{1/2} \tag{5.23}$$

5.6 Conclusion

This chapter gave a methodology on how to integrate the proposed skyhook control in a full car model to improve ride comfort and handling via a semi-active suspension system. It demonstrated the direct tilt control strategy which is able to tilt a vehicle body inwards. A technique to determine the vehicle rollover propensity to avoid tipping over has also been described. The road profile and four driving scenarios have been discussed in this chapter briefly which form a basis for the analysis described in the next two chapters. To compare and evaluate the results of the simulation environment and experimental setup, three criteria have been set. The ride comfort of the vehicle will be evaluated by comparing the results of the normalized RMS value of vehicle body vertical acceleration, vehicle pitch angular acceleration, roll angular acceleration and lateral acceleration. Evaluation of road handling performance will be done by comparing the tyre forces normalized to the static forces which acts on the wheels. A method to figure out the admissible acceleration level based on ISO 2631 has also been discussed in this chapter. The next chapter will contain the simulation results of the semi-active suspension system developed and described in this chapter.

Chapter 6
Simulation of Full Car Model

Abstract In this chapter, the analysis of the simulation results of the dynamic model of a full car model which considers the road bank angle is presented. The first section describes the parameters of the full car that have been used in the analysis model and the environment of the simulation. The second section describes the performance of the proposed skyhook control system under different road conditions. The performance of the combined approach, the proposed skyhook controller, is activated along with the direct tilt control which was evaluated in different driving scenarios in the third section. The next section is comprised of the summary of the simulation while the vehicle is travelling on road class C and following driving scenario four.

6.1 Simulation Environment

One objective of the suspension control system design in this manuscript is to tilt the vehicle inward when it is steered. The proposed skyhook control and the tilting mode of the direct tilt control are evaluated by the MATLAB and Simulink simulation of the full vehicle model. The road profile input has been assumed to be symmetrical for both left and right side. There is a time delay between the front and rear wheel which is calculated in real time according to the vehicle speed. At first, the proposed skyhook controlled semi-active suspension system has been activated alone to evaluate the controller performance. Then a simulation has been carried out to investigate the direct tilt control technique along with the proposed skyhook control system. Each system has been compared to the full vehicle passive suspension system to investigate the vehicle's ride handling and ride comfort performance. Three types of simulations were carried out: frequency domain and time domain simulations of vehicle ride comfort, road handling performance evaluation under different road conditions and vehicle tilting simulations under typical driving scenarios. Simulation parameters are listed in Table 6.1 (Fig. 6.1).

S. Kashem et al., *Vehicle Suspension Systems and Electromagnetic Dampers*,
Springer Tracts in Mechanical Engineering, DOI 10.1007/978-981-10-5478-5_6

Table 6.1 Nominal parameter values used in the simulation

Parameter	Definition	Values
m_1, m_4	Front left and right sprung mass	30.2 (kg)
m_2, m_3	Rear left and right sprung mass	49.7 (kg)
m	Sprung mass	809 (kg)
M_t	Total mass of the vehicle	968.8 (kg)
k_{i1} $(i = 1, 2, 3, 4)$	Tyre spring coefficient	181,000 (N/m)
k_{i2} $(i = 1, 2, 3, 4)$	Suspension spring coefficient	32,500 (N/m)
$c_1 \sim c_4$	Suspension damping coefficient	1400 (N-s/m)
F_{ri} $(i = 1, 2, 3, 4)$	Friction of suspension	300 (N)
a	Distance from CG to front axle	1.232 (m)
b	Distance from CG to rear axle	1.232 (m)
c, d	Half of wheel base	0.656 (m)
h	Height of CG of the vehicle	0.45 (m)
$C_{\alpha f}$	Cornering stiffness of front tyre	76,339 (N/rad)
$C_{\alpha r}$	Cornering stiffness of rear tyre	70,351 (N/rad)
I_x	Roll moment of inertial	330.5 (kg m^2)
I_y	Pitch moment of inertial	861.8 (kg m^2)
I_z	Yaw moment of inertial	2500 (kg m^2)
g	Gravity acceleration	9.8 (m/s^2)

6.2 Simulation with the Proposed Skyhook Controller

In order to evaluate the performance of the full car model with the proposed skyhook control, the simulation of the model needs to be carried out in relation to road surfaces that are defined by ISO to test the vehicle performance. In this simulation, the vehicle experienced road classes A, B and C defined by ISO 8608 and also combined road profile described in Sect. 4.2. All the simulations were run for 10 s except the simulation on combined road profile. The vehicle is travelling at a constant speed of 10 m/s on a straight road.

6.2.1 *Simulation on Road Class A*

In this section, the performance of the proposed skyhook controller on road class A is evaluated in the frequency domain and time domain separately.

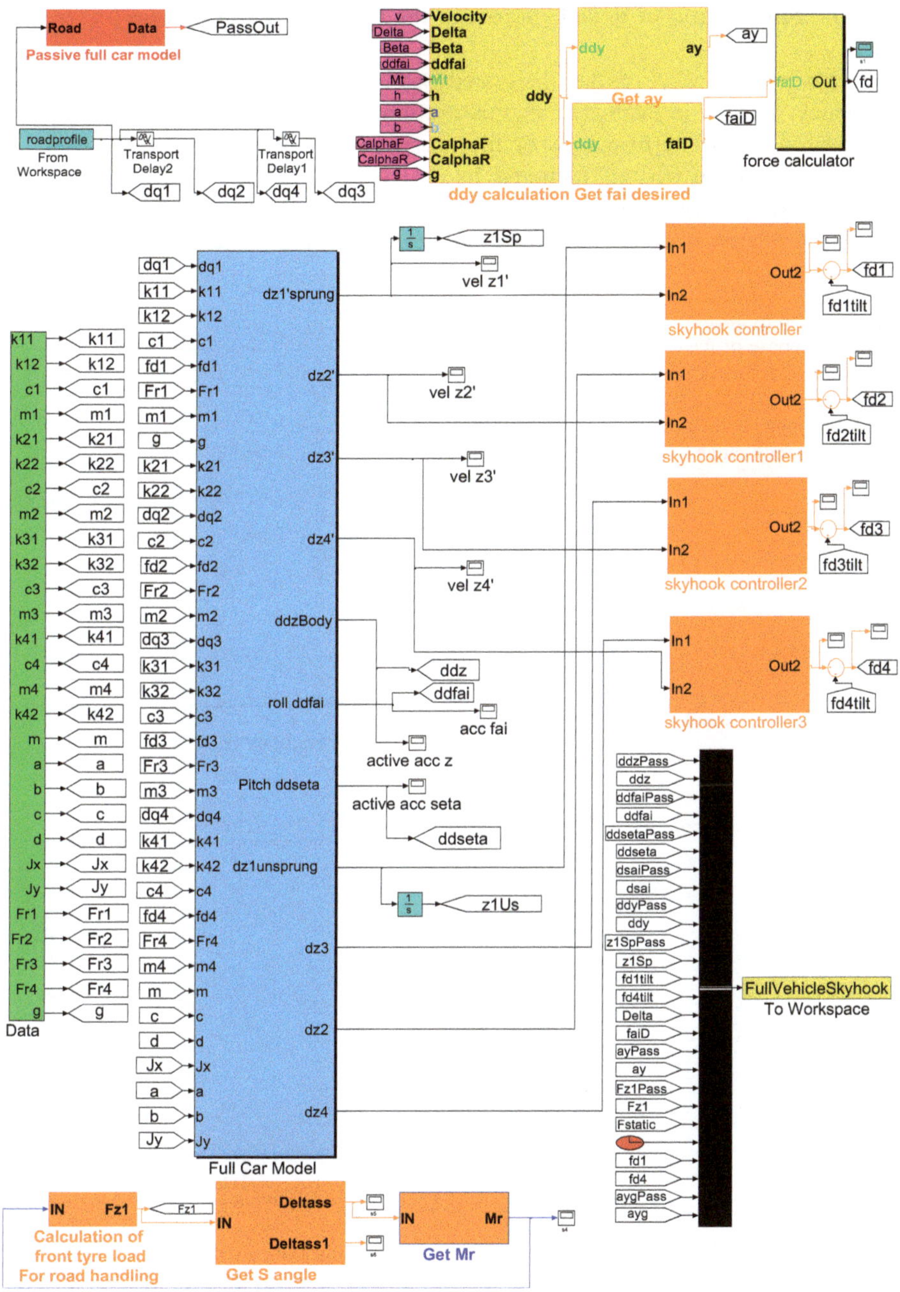

Fig. 6.1 Simulink model

6.2.1.1 Simulations of Ride Comfort in the Frequency Domain

The ride performance of the integrated semi-active suspension system in the frequency domain is evaluated in this section. For a convincing assessment, a comparison of the performance of the proposed skyhook control (SK) and the passive suspension system (denoted by PS) is presented. Figures 6.2 and 6.3 show the frequency responses of vehicle body vertical acceleration and pitch angular acceleration to the road disturbance class A. For better visualization, the

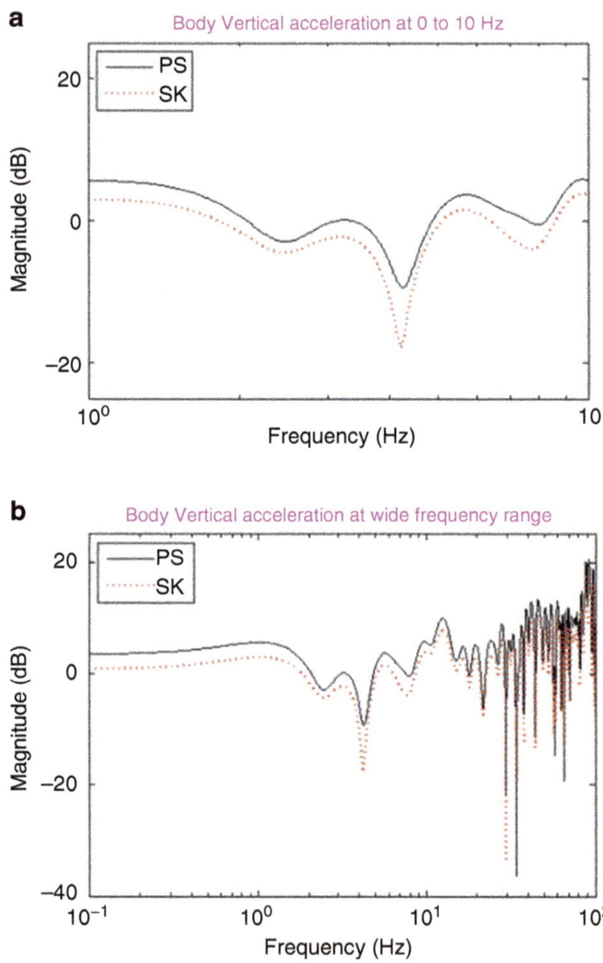

Fig. 6.2 The frequency domain response of the car body vertical acceleration to road class A: (**a**) at narrow frequency range and (**b**) at broad frequency range

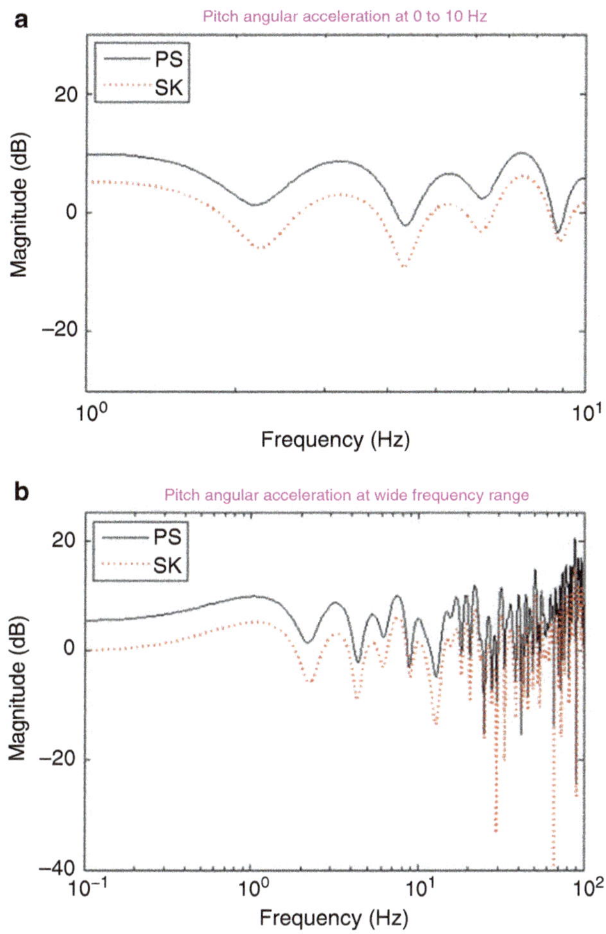

Fig. 6.3 The frequency domain response of the car body pitch angular acceleration to road class A: (**a**) at narrow frequency range and (**b**) at broad frequency range

responses of vehicle body vertical acceleration and pitch angular accelerations at 0–10 Hz have been plotted separately in a short time span in Figs. 6.2a and 6.3a, respectively. The study shows that human beings are very sensitive to vertical motion in the frequency range of 4–8 Hz [134]. So while designing a vehicle suspension system, sprung mass acceleration tolerance should be kept to a minimum in the frequency range mentioned above. From Figs. 6.2b and 6.3b, it is quite evident that the vehicle body vertical acceleration and pitch angular acceleration of the SK system remain closer to 0 db than the PS system at a wide range of frequencies, particularly in the region where humans are more sensitive to vibrations (Figs. 6.2a and 6.3a).

6.2.1.2 Simulations of Ride Comfort in the Time Domain

In this section, vehicle reaction to the road profile input by the passive and the proposed skyhook semi-active suspension systems is described. Figures 6.4, 6.5, and 6.6 show the time domain responses of vehicle vertical and pitch angular accelerations and the displacement of sprung mass m_1 to road disturbance class A. For better visualization, the response of vehicle vertical and pitch angular acceleration and the vertical displacement of sprung mass have been plotted separately in a short time span in Figs. 6.4b, 6.5b, and 6.6b, respectively. From Figs. 6.4a, 6.5a, and 6.6a, it is quite apparent that the vehicle vertical and pitch angular acceleration and the vertical displacement of the vehicle front left sprung mass of the SK system are better than the PS system. Compared to the passive suspension system, the NV (normalized body vertical acceleration) and NP

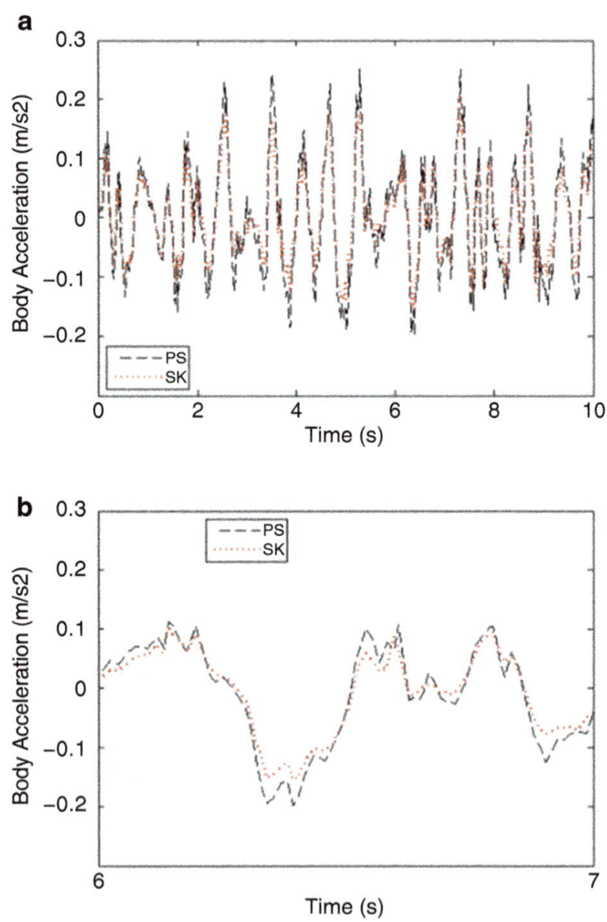

Fig. 6.4 The time domain response of vehicle body vertical acceleration to road class A: (**a**) full trajectory and (**b**) short time span

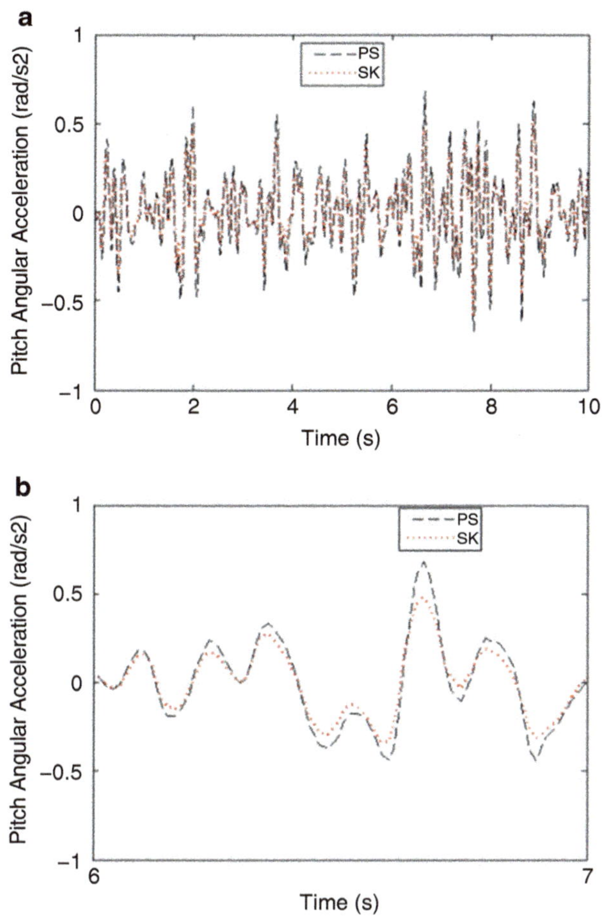

Fig. 6.5 The time domain response of vehicle pitch angular acceleration to road class A: (**a**) full trajectory and (**b**) short time span

(normalized pitch angular acceleration) of the skyhook controlled suspension system have decreased by 18.52% and 14.03%, respectively.

However, the road handling the performance of the SK controlled suspension system has decreased by only 1.07% compared to the passive suspension system. But the SK system's weighted RMS acceleration value of the sprung mass obtained in this simulation is 0.69 which is within the acceptable range of international code ISO 2631 [143]. Since the vehicle is traversing on a straight road, the differences in roll angular and lateral acceleration of both systems are negligible. Hence, the skyhook controlled suspension system is able to improve ride comfort of the vehicle effectively while the vehicle is moving on road class A.

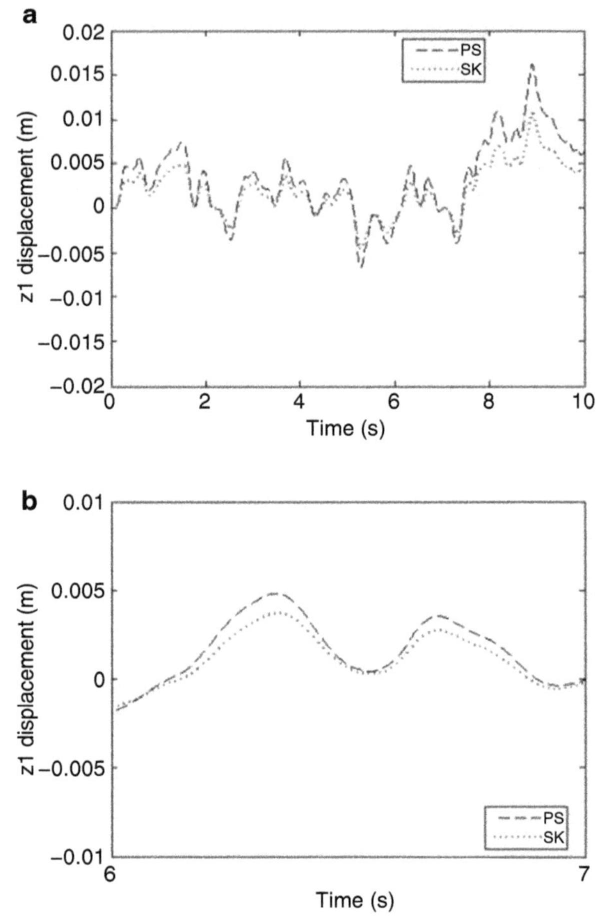

Fig. 6.6 The time domain response of vehicle pitch angular acceleration to road class A: (**a**) full trajectory and (**b**) short time span

6.2.2 *Simulation on Road Class B*

In this section, the performance of the proposed skyhook controller on road class B is evaluated in the frequency domain and time domain separately.

6.2.2.1 Simulations of Ride Comfort in the Frequency Domain

The ride performance of the integrated semi-active suspension system in the frequency domain is evaluated in this section. The responses of the vehicle body vertical acceleration and pitch angular acceleration at 0–10 Hz have been plotted separately in a short time span in Figs. 6.7a and 6.8a, respectively.

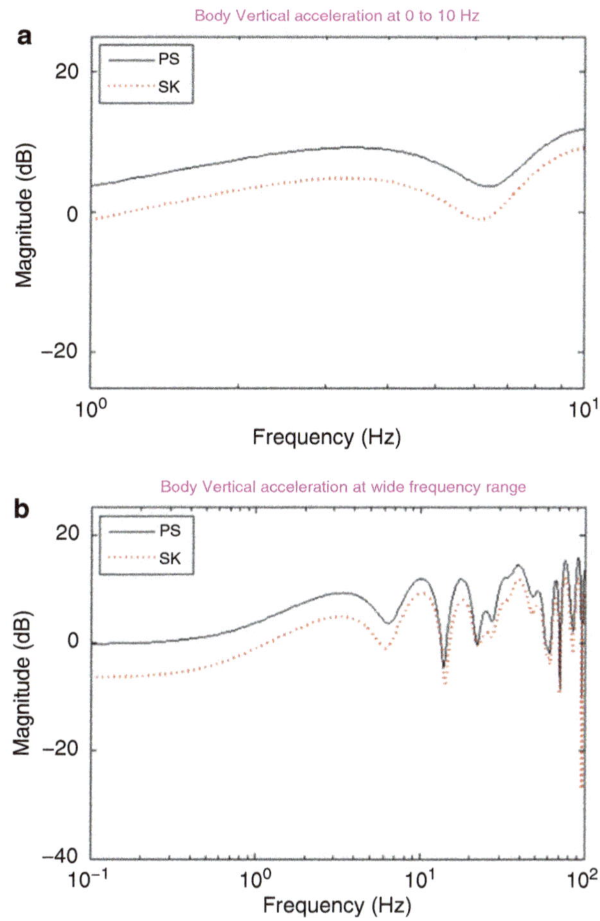

Fig. 6.7 The frequency domain response of the car body vertical acceleration to road class B: (**a**) at narrow frequency range and (**b**) at broad frequency range

From Figs. 6.7b and 6.8b, it is quite evident that the body vertical acceleration and pitch angular acceleration of the SK system remain closer to 0 db than the PS system at a wide range of frequencies, particularly in the region where humans are more sensitive to vibrations (Figs. 6.7a and 6.8a).

6.2.2.2 Simulations of Ride Comfort in the Time Domain

In this section, vehicle reactions to the road profile input by the passive and the proposed skyhook semi-active suspension systems are described. For better visualization, the responses of vehicle vertical and pitch angular acceleration and the vertical displacement of sprung mass have been plotted separately in a short time span in Figs. 6.9b, 6.10b, and 6.11b, respectively.

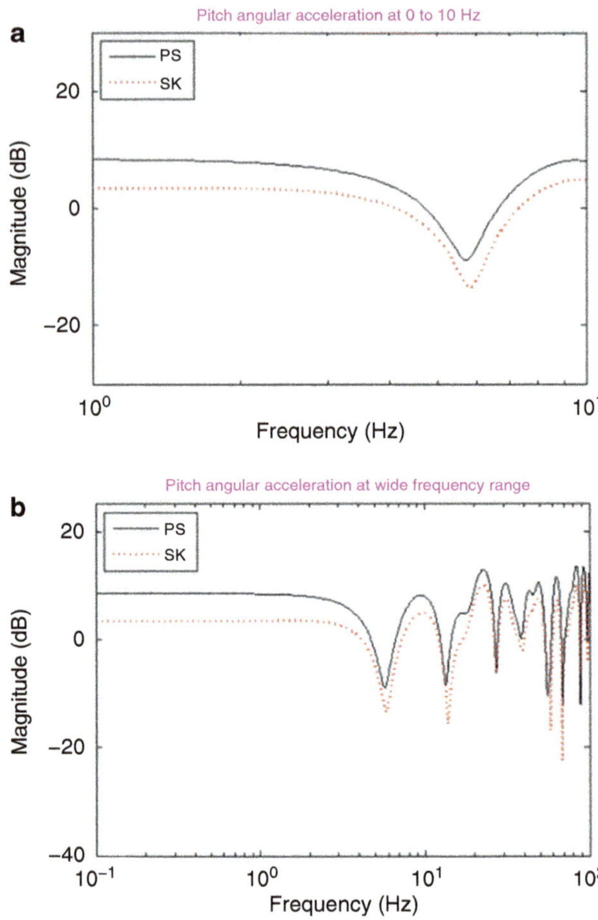

Fig. 6.8 The frequency domain response of the car body pitch angular acceleration to road class B: (**a**) at low frequency and (**b**) at broad frequency range

From Figs. 6.9a, 6.10a, and 6.11a, it is quite apparent that the vehicle vertical and pitch angular accelerations and the vertical displacement of the vehicle front left sprung mass of the SK system are better than the PS system. Compared to the passive suspension system, the NV (normalized body vertical acceleration) and NP (normalized pitch angular acceleration) of the skyhook controlled suspension system have decreased by 18.02% and 14.53%, respectively.

However, the road handling performance of the SK controlled suspension system decreased by only 1.08% compared to the passive suspension system. But the SK system's weighted RMS acceleration value of the sprung mass obtained in this simulation is 0.71 which is within the acceptable range of international code ISO 2631 [143]. Since the vehicle is traversing on a straight road, the differences in roll angular and lateral acceleration of both systems are negligible. Hence, the

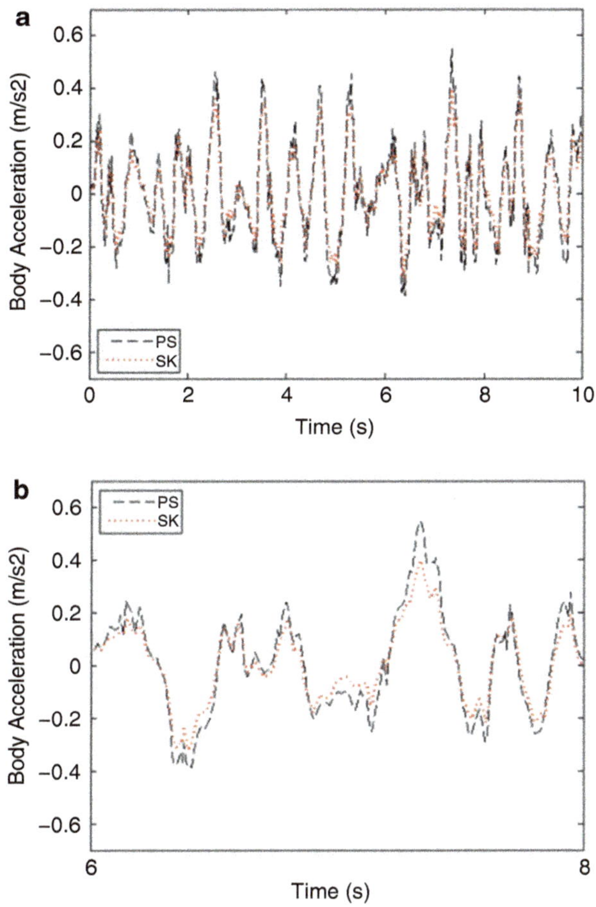

Fig. 6.9 The time domain response of vehicle body vertical acceleration to road class B: (**a**) full trajectory and (**b**) short time span

skyhook controlled suspension system is able to improve ride comfort of the vehicle effectively while the vehicle is moving on road class B.

6.2.3 Simulation on Road Class C

In this section, the performance of the proposed skyhook controller on road class C is evaluated in the frequency domain and time domain separately.

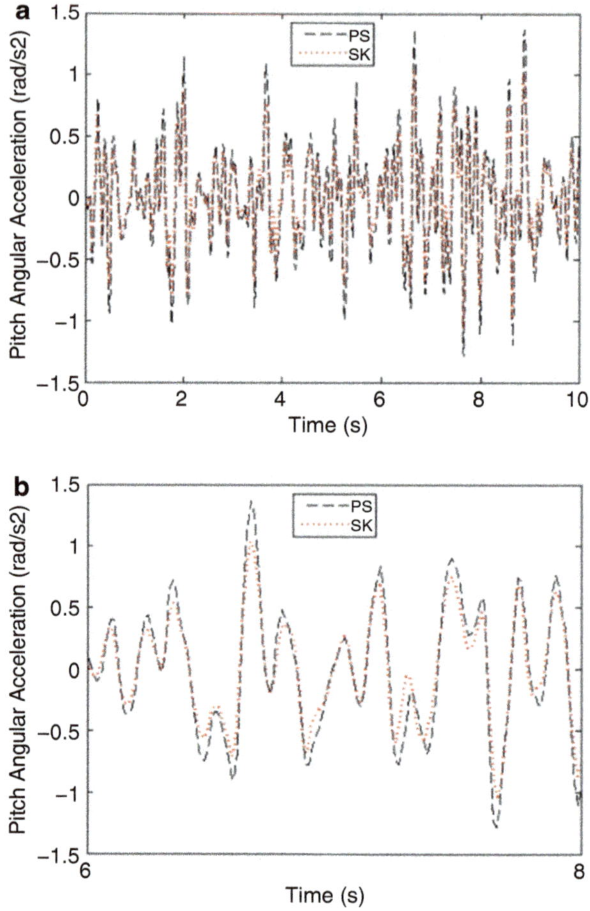

Fig. 6.10 The time domain response of vehicle pitch angular acceleration to road class B: (**a**) full trajectory and (**b**) short time span

6.2.3.1 Simulations of Ride Comfort in the Frequency Domain

The ride performance of the integrated semi-active suspension system in the frequency domain is evaluated in this section. The responses of vehicle body vertical acceleration and pitch angular acceleration at 0–10 Hz have been plotted separately in a short time span in Figs. 6.12a and 6.13a, respectively.

From Figs. 6.12b and 6.13b, it is quite evident that the vehicle body vertical acceleration and pitch angular acceleration of the SK system remain closer to 0 db than the PS system at a wide range of frequencies, particularly in the region where humans are more sensitive to vibrations (Figs. 6.12a and 6.13a).

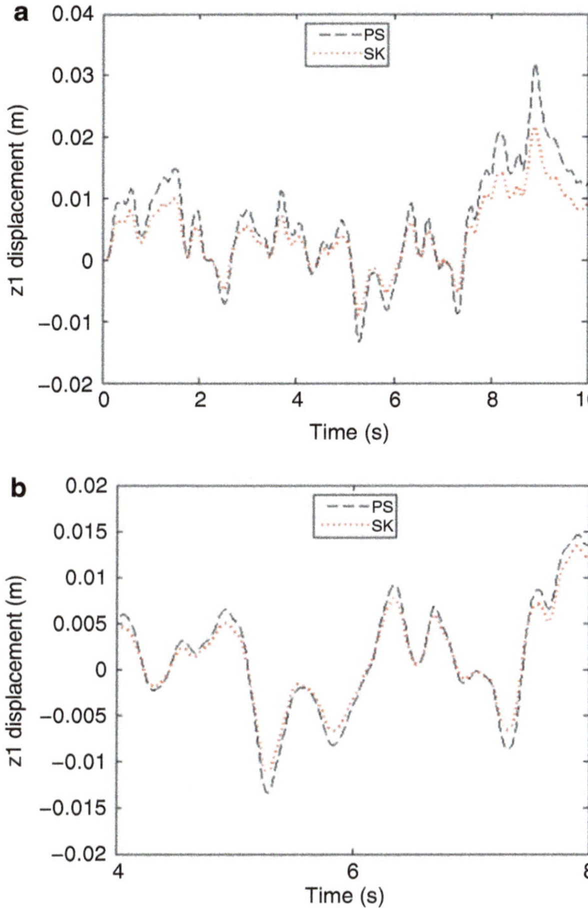

Fig. 6.11 The time domain response of the vehicle sprung mass m_1 vertical displacement to road class B: (**a**) full trajectory and (**b**) short time span

6.2.3.2 Simulations of Ride Comfort in the Time Domain

In this section, vehicle reaction to the road profile input by the passive and the proposed skyhook semi-active suspension systems is described. For better visualization, the responses of vehicle vertical and pitch angular acceleration and the vertical displacement of sprung mass have been plotted separately in a short time span in Figs. 6.14b, 6.15b, and 6.16b, respectively. From Figs. 6.14a, 6.15a, and 6.16a, it is quite apparent that the vehicle vertical and pitch angular accelerations and the vertical displacement of the vehicle front left sprung mass of the SK system are better than the PS system. Compared to the passive suspension system, the NV (normalized body vertical acceleration) and NP (normalized pitch angular acceleration) of the skyhook controlled suspension system have decreased by 18.32% and 14.92%, respectively.

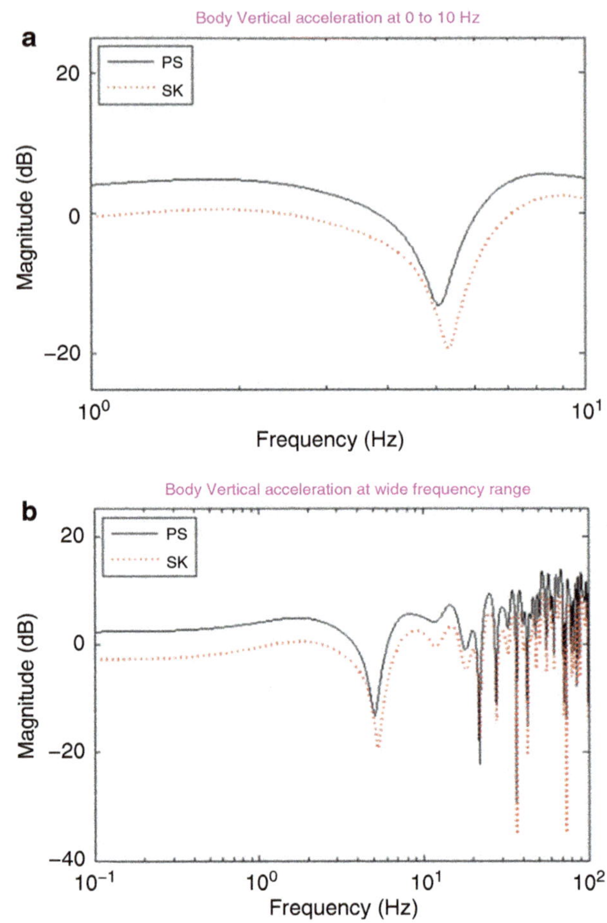

Fig. 6.12 The frequency domain response of the car body vertical acceleration to road class C: (**a**) at narrow frequency range and (**b**) at broad frequency range

However, the road handling performance of the SK controlled suspension system has decreased by only 1.68% compared to the passive suspension system. But the SK system's weighted RMS acceleration value of the sprung mass obtained in this simulation is 0.72 which is within the acceptable range of international code ISO 2631 [143]. Since the vehicle is traversing on a straight road, the differences in roll angular and lateral acceleration of both systems are negligible. Hence, the skyhook controlled suspension system is able to improve ride comfort of the vehicle effectively while the vehicle is moving on road class C.

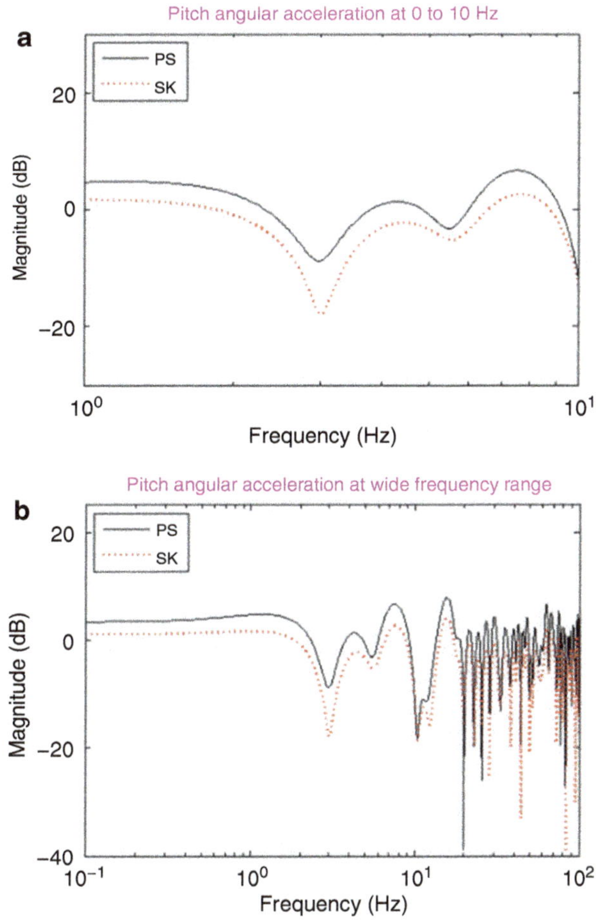

Fig. 6.13 The frequency domain response of the car body pitch angular acceleration to road class C: (**a**) at narrow frequency range and (**b**) at broad frequency range

6.2.4 Simulation on Combined Road

In this section, the performance of the proposed skyhook controller on a combined road surface (generated using the three classes A, B and C of road profiles where each class lasts for 20 s) is evaluated in the frequency domain and time domain separately.

6.2.4.1 Simulations of Ride Comfort in the Frequency Domain

The ride performance of the integrated semi-active suspension system in the frequency domain is evaluated in this section. The responses of the vehicle body

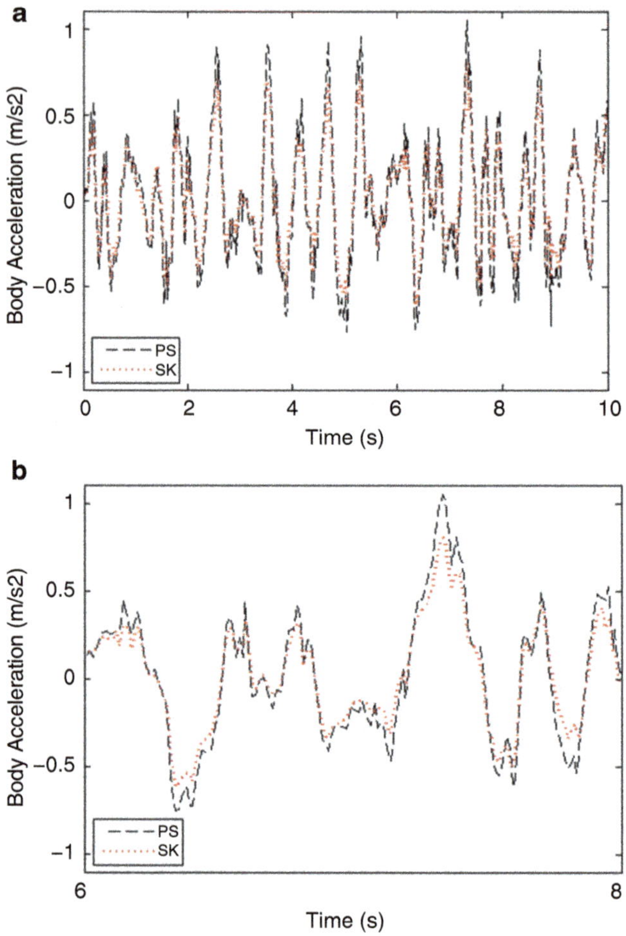

Fig. 6.14 The time domain response of vehicle body vertical acceleration to road class C: (**a**) full trajectory and (**b**) short time span

vertical acceleration and pitch angular acceleration at 0–10 Hz have been plotted separately in a short time span in Figs. 6.17a and 6.18a, respectively.

From Figs. 6.17b and 6.18b, it is quite evident that the vehicle body vertical acceleration and pitch angular acceleration of the SK system remain closer to 0 db than the PS system at a wide range of frequencies, particularly in the region where humans are more sensitive to vibrations (Figs. 6.17a and 6.18a).

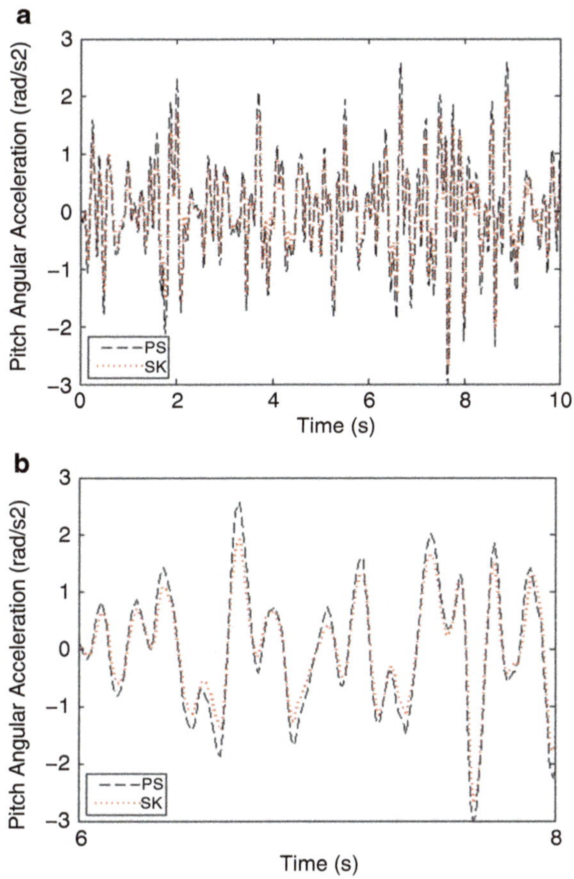

Fig. 6.15 The time domain response of vehicle pitch angular acceleration to road class C: (**a**) full trajectory and (**b**) short time span

6.2.4.2 Simulations of Ride Comfort in the Time Domain

In this section, vehicle reactions to the road profile input by the passive and the proposed skyhook semi-active suspension systems are described. For better visualization, the responses of vehicle vertical and pitch angular acceleration and the vertical displacement of sprung mass have been plotted separately in a short time span in Figs. 6.19b, 6.20b, and 6.21b, respectively. From Figs. 6.19a, 6.20a, and 6.21a, it is quite apparent that the vehicle vertical and pitch angular accelerations and the vertical displacement of the vehicle front left sprung mass of the SK system are better than the PS system. Compared to the passive suspension system, the NV (normalized body vertical acceleration) and NP (normalized pitch angular acceleration) of the skyhook controlled suspension system have decreased by 18.92% and 14.73%, respectively.

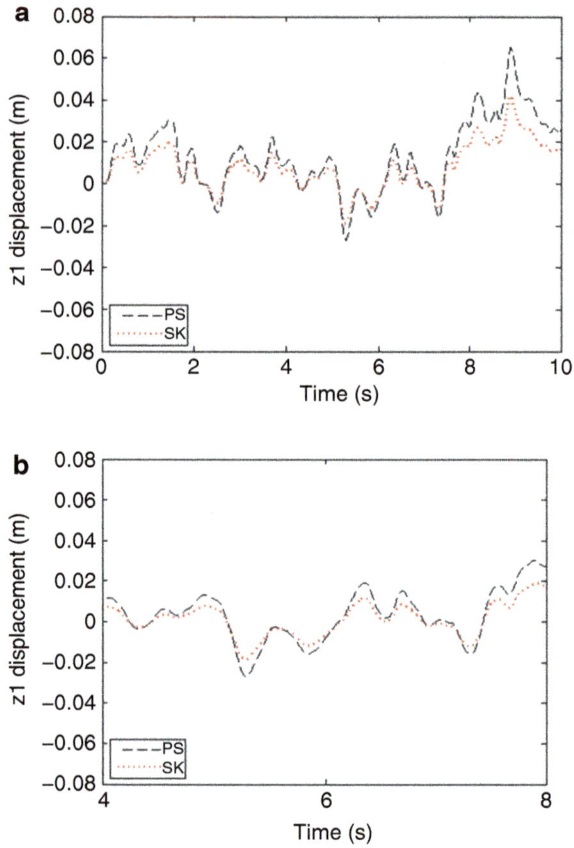

Fig. 6.16 The time domain response of the vehicle sprung mass m_1 vertical displacement to road class C: (**a**) full trajectory and (**b**) short time span

However, the road handling performance of the SK controlled suspension system has decreased by only 2.08% compared to the passive suspension system. But the SK system's weighted RMS acceleration value of the sprung mass obtained in this simulation is 0.79 which is within the acceptable range of international code ISO 2631 [143]. Since the vehicle is traversing on a straight road, the differences in roll angular and lateral acceleration of both systems are negligible. Hence, the skyhook controlled suspension system is able to improve ride comfort of the vehicle effectively while the vehicle is moving on combined road surface.

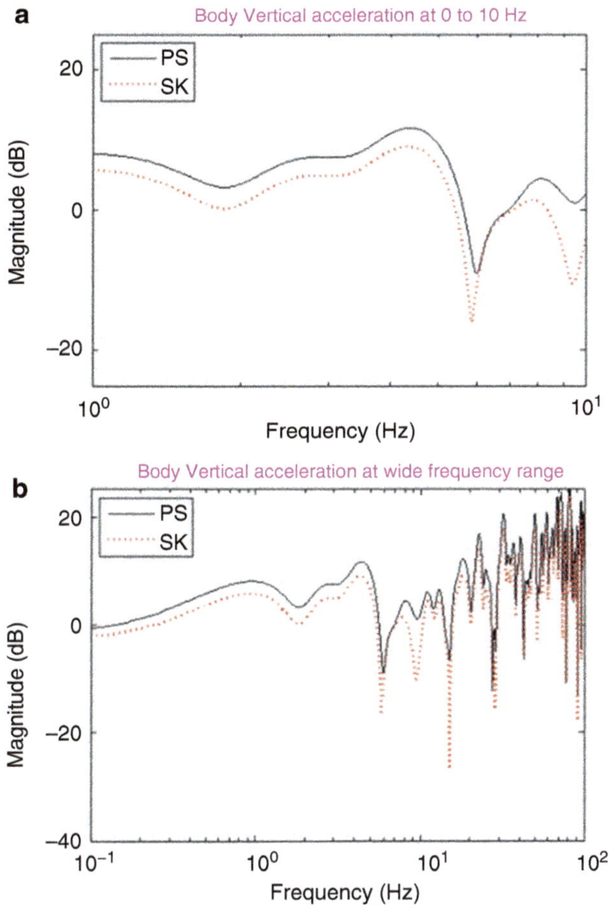

Fig. 6.17 The frequency domain response of the car body vertical acceleration to the combined road: (**a**) at narrow frequency range and (**b**) at broad frequency range

6.3 Simulation with Skyhook and Direct Tilt Controller

The objective of this section is to study the response of the vehicle when it is tilted due to steering angle and road bank angle. The tilting mode of the system was evaluated using the MATLAB and Simulink simulation where the direct tilt control is activated along with the proposed skyhook controller.

In this simulation, the vehicle experienced road class C defined by ISO 8608 described in Sect. 4.2. The vehicle is travelling at the same speed of 10 m/s. The vehicle's trajectory is set by four driving scenarios depicted in Sect. 5.4 and the response of the vehicle travelling on road class C with each scenario described below.

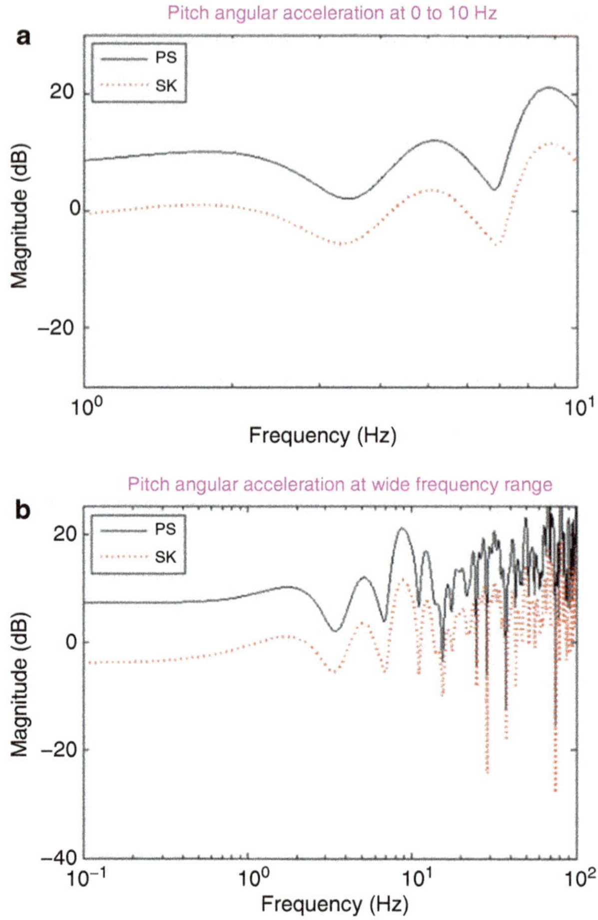

Fig. 6.18 The frequency domain response of the car body pitch angular acceleration to the combined road: (**a**) at narrow frequency range and (**b**) at broad frequency range

6.3.1 Simulation on Driving Scenario One

In this section, vehicle reaction to driving scenario one using the PS and the SKDT semi-active suspension system is described. Figure 6.22a shows the response of the vehicle's desired tilting angle (faiD) for the steering input signal (Delta) and the road bank angle (β) according to the designed direct tilt control system. The required actuator force to tilt the vehicle according to the desired tilting angle is depicted in Fig. 6.43b.

The time domain responses of the vehicle body vertical, pitch angular, roll angular and lateral acceleration and the vertical displacement of the vehicle front left sprung mass m_1 for this driving scenario have been shown in Figs. 6.23, 6.24, 6.25, 6.26, and 6.27, respectively, and it is quite evident that the displacement of sprung mass m_1, the vehicle vertical and pitch angular accelerations and the lateral acceleration of the SKDT system have decreased compared to the PS system. For

Fig. 6.19 The time domain response of vehicle body vertical acceleration to the combined road: (a) full trajectory and (b) short time span

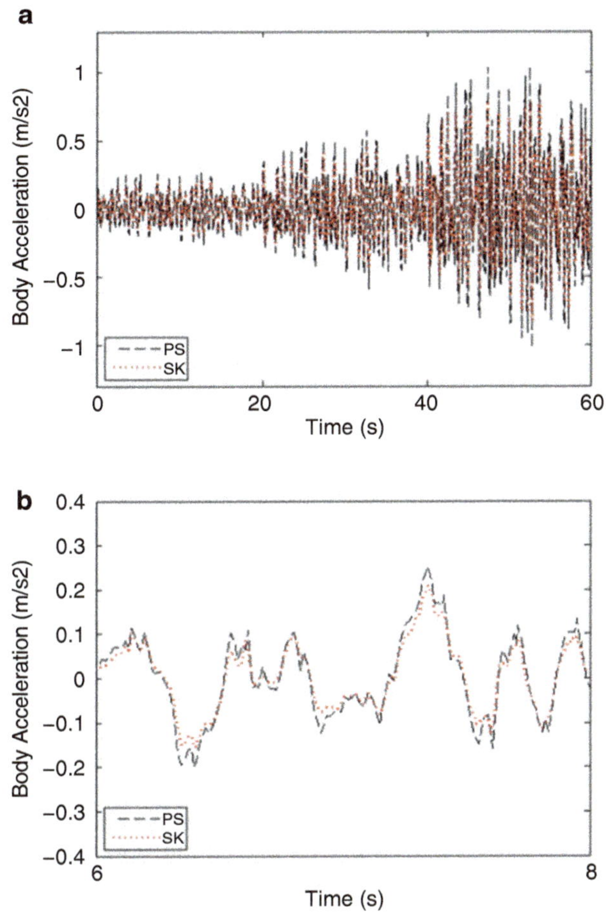

better visualization, the response of the displacement of sprung mass m_1, vehicle vertical and pitch angular roll angular and lateral acceleration and the maximum lateral acceleration value over gravity (a_y/g) have been plotted separately in a short time span.

Compared to the passive suspension system, the NV and NP of the SKDT suspension system have decreased by 13% and 11%, respectively. However, the normalized RMS of the roll angular acceleration (NR) of the SKDT is increased by 10%. Fortunately, the vehicle's roll angular acceleration only accounts for less than 20% of vehicle ride comfort, in terms of the ride comfort index specified by the ISO 2631-1 standard. Hence, the vehicle ride comfort would not be affected significantly by the large roll angular acceleration. Moreover, the passengers would feel less lateral acceleration; hence, the ride comfort would be improved, as shown in Fig. 6.26. Compared to the PS system, the NL (normalized lateral acceleration) of the SKDT suspension system is decreased by 10.01%.

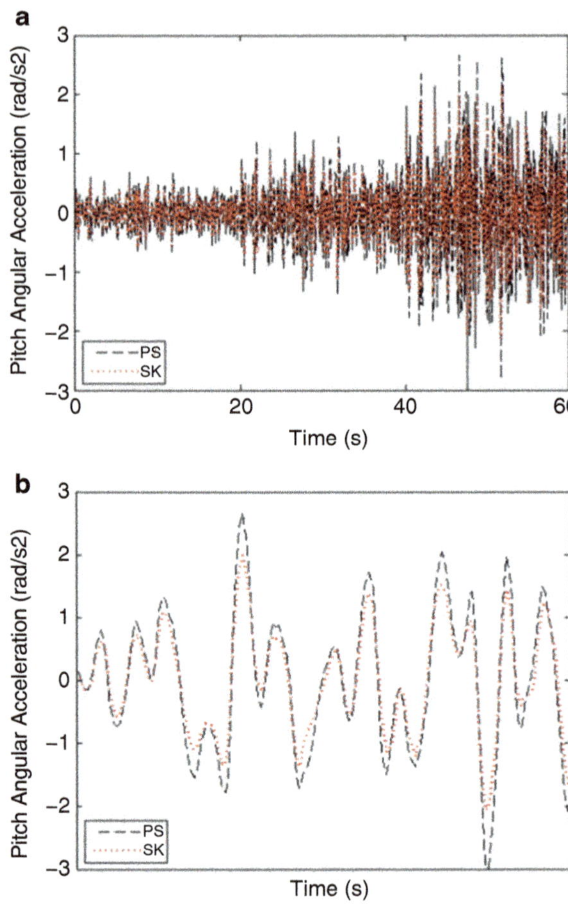

Fig. 6.20 The time domain response of vehicle pitch angular acceleration to the combined road: (**a**) full trajectory and (**b**) short time span

If the body of the vehicle is parallel with the ground surface, its rollover threshold becomes a function of the vehicle's track width (t) and the centre of gravity height (h) as depicted in Eq. (5.17). The maximum allowable rollover threshold of this vehicle is 1.45. The rollover threshold defines the maximum lateral acceleration value over gravity (a_y/g) that the vehicle would be able to reach before tipping over. From Fig. 6.28a, b, it is apparent that (a_y/g) never exceeds the maximum allowable rollover threshold of this vehicle. Hence, the vehicle remains stable throughout the trajectory of this driving scenario.

However, the road handling performance of the skyhook controlled suspension system has decreased by only 2.3% compared to the passive suspension system. But the SKDT system's weighted RMS acceleration value of the sprung mass obtained in this simulation is 0.71 which is within the acceptable range. Hence, the integrated suspension system is able to improve ride comfort of the vehicle effectively while the vehicle is moving on road class C with this driving scenario.

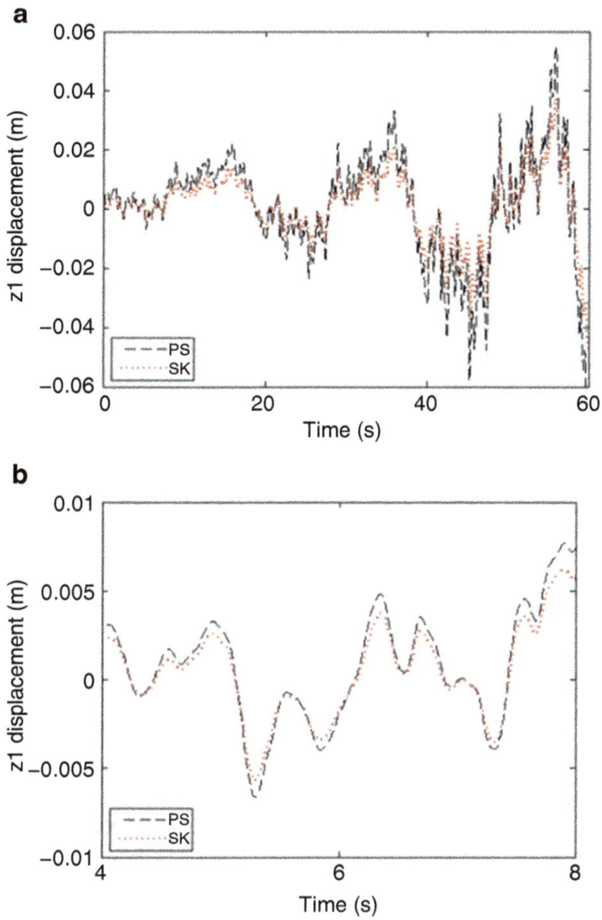

Fig. 6.21 The time domain response of the vehicle sprung mass m_1 vertical displacement to the combined road: (**a**) full trajectory and (**b**) short time span

6.3.2 Simulation on Driving Scenario Two

In this section, vehicle reaction to driving scenario two using the PS and the SKDT semi-active suspension system is described. Figure 6.29a shows the response of the vehicle's desired tilting angle for the steering input signal and the road bank angle according to the designed direct tilt control system. The required actuator force to tilt the vehicle according to the desired tilting angle is depicted in Fig. 6.29b.

The time domain responses of the vertical displacement of the vehicle front left sprung mass m_1, vehicle body vertical, pitch angular, roll angular and lateral acceleration for this driving scenario are shown in Figs. 6.30, 6.31, 6.32, 6.33, and 6.34, respectively, and it is quite evident that the displacement of sprung mass m_1, the vehicle vertical and pitch angular accelerations and the lateral acceleration of the SKDT system have decreased compared to the PS system. For better

Fig. 6.22 The response of steering and bank angle in driving scenario one: (**a**) desired tilting angle, (**b**) required actuator force

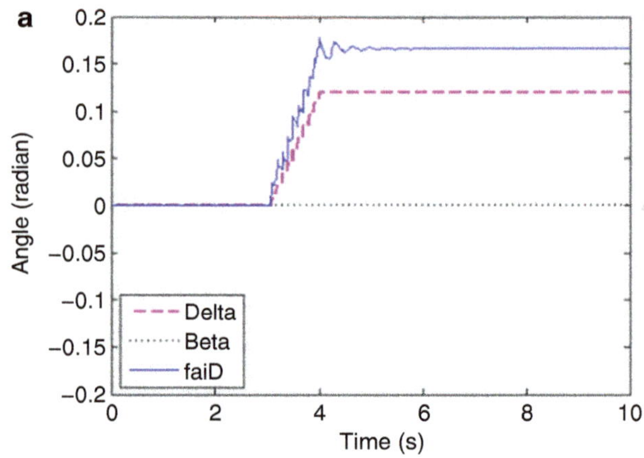

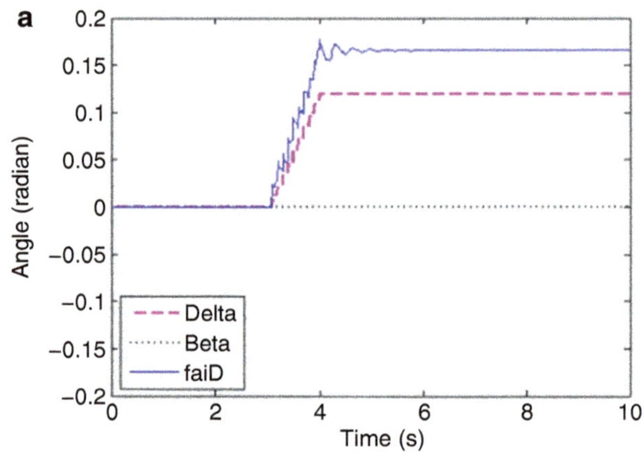

visualization, the responses of the displacement of sprung mass m_1, vehicle vertical, pitch angular roll angular and lateral acceleration and the maximum lateral acceleration value over gravity (a_y/g) have been plotted separately in a short time span.

Compared to the passive suspension system, the NV and NP of the SKDT suspension system have decreased by 12.96% and 10.94 %, respectively. However, the normalized RMS roll angular acceleration (NR) of the SKDT system has increased by 9.85%. Compared to the PS system, the NL (normalized lateral acceleration) of the SKDT suspension system has decreased by 10.11%.

If the body of the vehicle is parallel with the ground surface, its rollover threshold becomes a function of the vehicle's track width (t) and the centre of gravity height (h) as depicted in Eq. (5.17). The maximum allowable rollover threshold of this vehicle is 1.45. The rollover threshold defines the maximum lateral

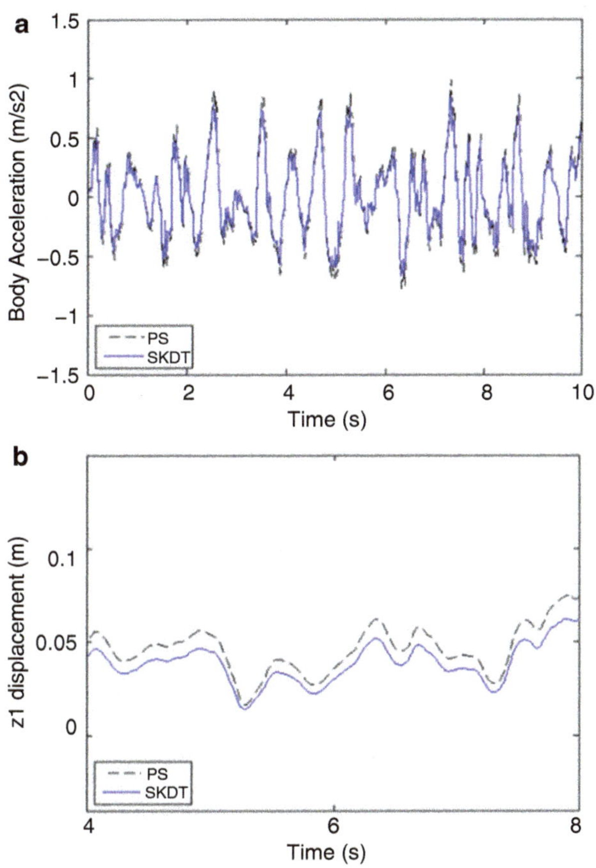

Fig. 6.23 The vehicle body vertical acceleration for driving scenario one: (**a**) full trajectory and (**b**) short time span

acceleration value over gravity (a_y/g) that the vehicle would be able to reach before tipping over. From Fig. 6.35a, b, it is apparent that (a_y/g) never exceeds the maximum allowable rollover threshold of this vehicle. Hence, the vehicle remains stable throughout the trajectory of this driving scenario.

However, the road handling performance of skyhook controlled suspension system has decreased by only 2.33% compared to the passive suspension system. But the SKDT system's weighted RMS acceleration value of the sprung mass obtained in this simulation is 0.76 which is within the acceptable range. Hence, the integrated suspension system is able to improve ride comfort of the vehicle effectively while the vehicle is moving on road class C with this driving scenario.

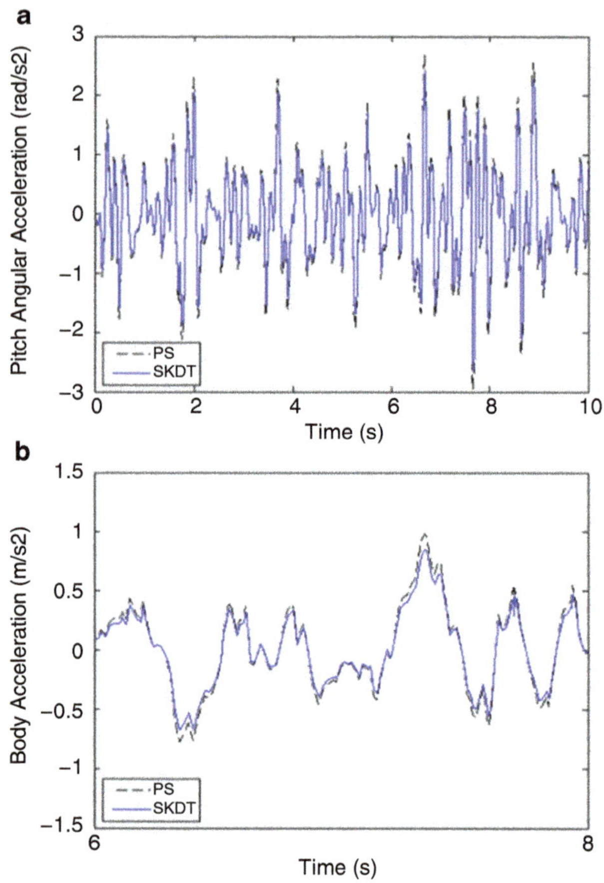

Fig. 6.24 The pitch angular acceleration for driving scenario one: (**a**) full trajectory and (**b**) short time span

6.3.3 Simulation on Driving Scenario Three

In this section, vehicle reaction to the driving scenario using the PS and the SKDT semi-active suspension systems is described. Figure 6.36a shows the response of the faiD for the input signal delta and the β according to the designed direct tilt control. The required actuator force to tilt the vehicle according to the desired tilting angle is depicted in Fig. 6.36b.

The time domain responses of the vertical displacement of the vehicle front left sprung mass m_1, vehicle body vertical, pitch angular, roll angular and lateral acceleration for this driving scenario are shown in Figs. 6.37, 6.38, 6.39, 6.40, and 6.41, respectively, and it is quite evident that the displacement of sprung mass m_1, the vehicle vertical, pitch angular accelerations and the lateral acceleration of the SKDT system have decreased compared to the PS system. For better visualization, the responses of the displacement of sprung mass m_1, vehicle vertical, pitch angular roll angular and lateral acceleration and the maximum lateral

Fig. 6.25 The roll angular acceleration for driving scenario one: (**a**) full trajectory and (**b**) short time span

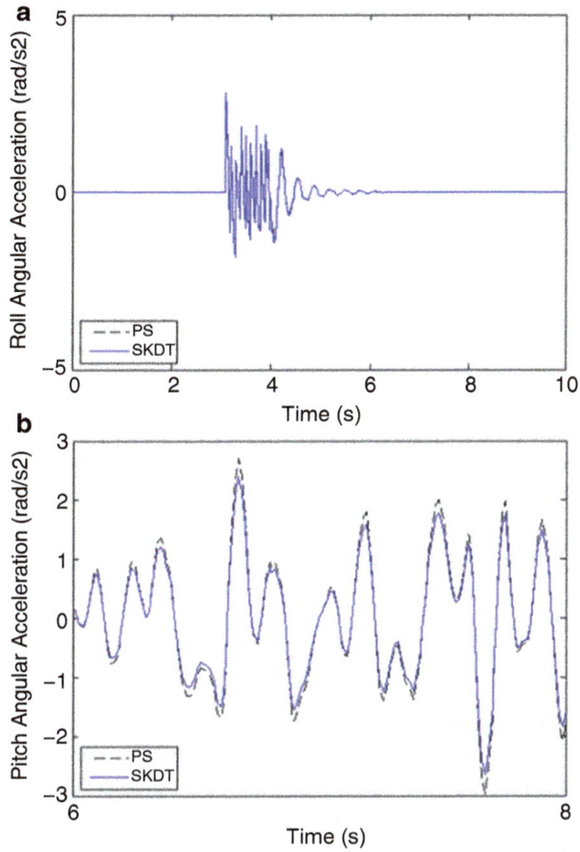

acceleration value over gravity (a_y/g) have been plotted separately in a short time span.

Compared to the passive suspension system, the NV and NP of the SKDT suspension system have decreased by 12.66% and 10.44 %, respectively. However, the normalized RMS roll angular acceleration (NR) of the SKDT system has increased by 9.95%. Compared to the PS system, the NL (normalized lateral acceleration) of the SKDT suspension system has decreased by 10.61%.

If the body of the vehicle is parallel with the ground surface, its rollover threshold becomes a function of the vehicle's track width (t) and the centre of gravity height (h) as depicted in Eq. (5.17). The maximum allowable rollover threshold of this vehicle is 1.45. The rollover threshold defines the maximum lateral acceleration value over gravity (a_y/g) that the vehicle would be able to reach before tipping over. From Fig. 6.42a, b, it is apparent that (a_y/g) never exceeds the maximum allowable rollover threshold of this vehicle. Hence, the vehicle remains stable throughout the trajectory of this driving scenario.

However, the road handling performance of the skyhook controlled suspension system has decreased by only 2.23% compared to the passive suspension system. But the SKDT system's weighted RMS acceleration value of the sprung mass obtained in this simulation is 0.69 which is within the acceptable range. Hence, the integrated suspension system is able to improve ride comfort of the vehicle effectively while the vehicle is moving on road class C with this driving scenario.

6.3.4 Simulation on Driving Scenario Four

In this section, vehicle reaction to the driving scenario using the PS and the SKDT semi-active suspension system is described. Figure 6.43a shows the response of the vehicle's desired tilting angle (faiD) for the steering input signal (delta) and the road bank angle (β) according to the designed direct tilt control system. The required actuator force to tilt the vehicle according to the desired tilting angle is depicted in Fig. 6.43b.

The time domain responses of the vertical displacement of the vehicle front left sprung mass m_1, vehicle body vertical, pitch angular, roll angular and lateral acceleration for this driving scenario are shown in Figs. 6.44, 6.45, 6.46, 6.47, and 6.48, respectively, and it is quite evident that the displacement of sprung mass m_1, the vehicle vertical, pitch angular accelerations and the lateral acceleration of the SKDT system have decreased compared to the PS system. For better visualization, the responses of the displacement of sprung mass m_1, vehicle vertical, pitch angular roll angular and lateral acceleration and the maximum lateral acceleration value over gravity (a_y/g) have been plotted separately in a short time span.

Compared to the passive suspension system, the NV and NP of the SKDT suspension system have decreased by 12.99% and 10.99 %, respectively. However, the normalized RMS roll angular acceleration (NR) of the SKDT system has increased by 9.99%. Fortunately, the vehicle's roll angular acceleration only accounts for less than 20% of vehicle ride comfort, in terms of ride comfort index specified by the ISO 2631-1 standard. Hence, the vehicle ride comfort would not be affected significantly by the large roll angular acceleration. Moreover, the passengers would feel less lateral acceleration; hence, the ride comfort would be improved, as shown in Fig. 6.48. Compared to the PS system, the NL (normalized lateral acceleration) of the SKDT suspension system has decreased by 10%.

If the body of the vehicle is parallel with the ground surface, its rollover threshold becomes a function of the vehicle's track width (t) and the centre of gravity height (h) as depicted in Eq. (5.17). The maximum allowable rollover threshold of this vehicle is 1.45. The rollover threshold defines the maximum lateral acceleration value over gravity (a_y/g) that the vehicle would be able to reach before tipping over.

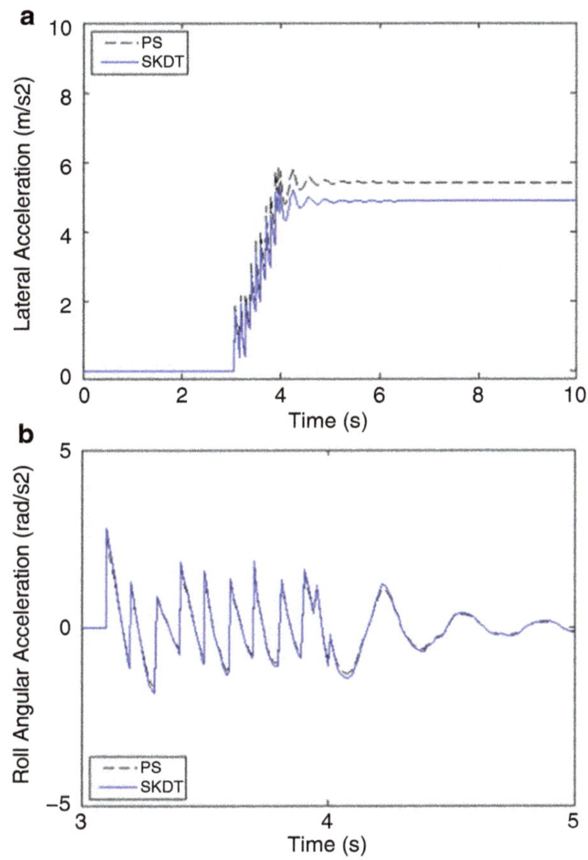

Fig. 6.26 The lateral acceleration for driving scenario one: (**a**) full trajectory and (**b**) short time span

From Fig. 6.49a, b, it is apparent that (a_y/g) never exceeds the maximum allowable rollover threshold of this vehicle. Hence, the vehicle remains stable throughout the trajectory of this driving scenario.

However, the road handling the performance of the skyhook controlled suspension system has decreased by only 2% compared to the passive suspension system. But the SKDT system's weighted RMS acceleration value of the sprung mass obtained in this simulation is 0.72 which is within the acceptable range.

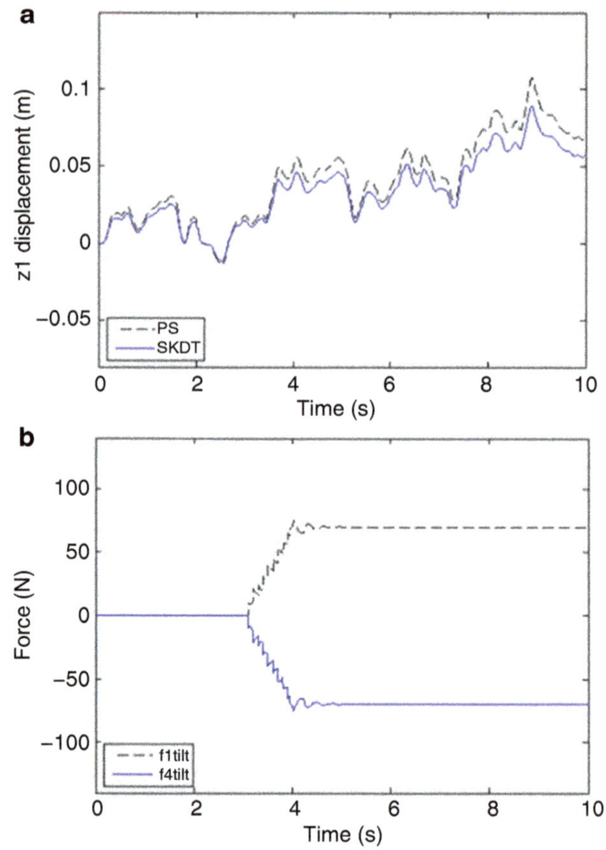

Fig. 6.27 The vehicle sprung mass m_1's vertical displacement for driving scenario one: (**a**) full trajectory and (**b**) short time span

6.4 Simulation Summary

The objective of this section is to summarize the evaluation of the SK system (the proposed skyhook suspension system) and the SKDT system where the direct tilt control is activated along with the proposed skyhook controller in the same simulation. The performances of the SK and SKDT systems are compared to the PS system (the passive suspension system). In this simulation, the vehicle experienced road class C defined by ISO 8608 described in Sect. 5.4. The vehicle trajectory defined by the driving scenario four (Sect. 5.4). This driving scenario has been used as if the vehicle experienced both left and right turns. The frequency and time domain analyses are presented separately in the following sections.

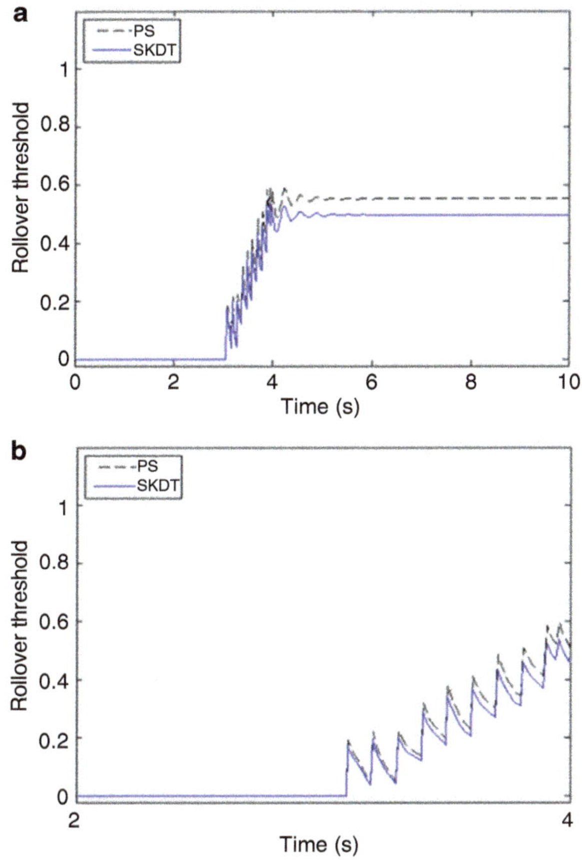

Fig. 6.28 The rollover threshold in driving scenario one: (**a**) full trajectory and (**b**) short time span

6.4.1 Simulations of Ride Comfort in the Frequency Domain

Figures 6.50 and 6.51 show the frequency responses of the vehicle vertical, pitch angular and roll angular acceleration to the road disturbance class C. For better visualization, the responses of vehicle vertical, pitch angular and roll angular accelerations at 0–10 Hz have been plotted separately in a short time span in Figs. 6.50a, 6.51a, and 6.52a, respectively.

The study shows that human beings are very sensitive to the vertical motion in the frequency range of 4–8 Hz [134]. So while designing a vehicle suspension system, sprung mass acceleration tolerance should be kept to a minimum in the frequency range mentioned above. From Figs. 6.50b and 6.51b, it is quite evident that the vehicle body vertical acceleration and pitch angular acceleration of the SK and SKDT systems remained closer to 0 db than the PS system at a wide range

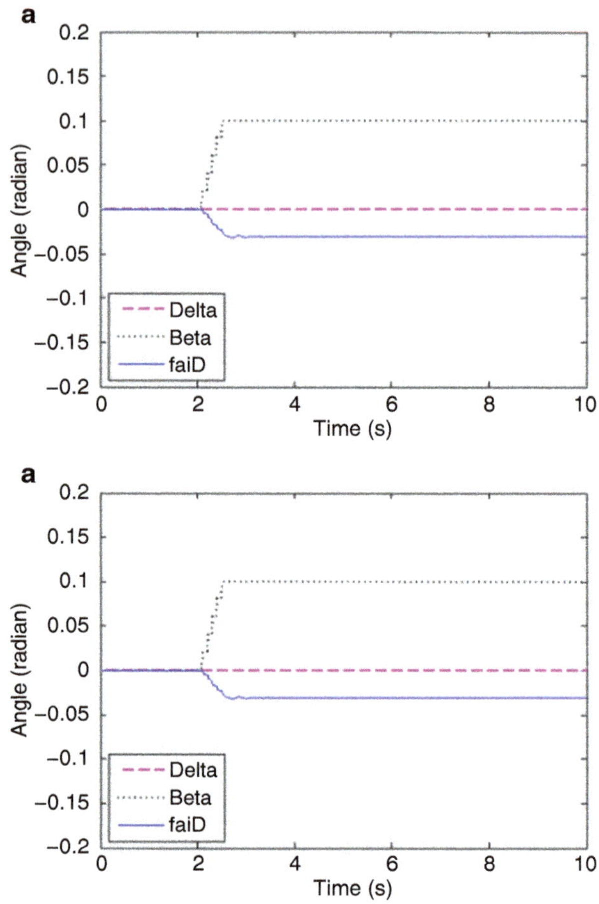

Fig. 6.29 The response of steering and bank angle in driving scenario two: (**a**) desired tilting angle, (**b**) required actuator force

of frequencies, especially at frequencies where humans are more sensitive to vibrations (Figs. 6.50a and 6.51a).

However, the frequency responses of roll angular acceleration of the SK and SKDT systems have deteriorated compared to the PS system as shown in Fig. 6.52. Fortunately, the vehicle's roll angular acceleration only accounts for less than 20% of vehicle ride comfort, in terms of ride comfort index specified by the ISO 2631-1 standard. Hence, the vehicle ride comfort would not be affected significantly by the large roll angular acceleration. Moreover, the passengers would feel less lateral acceleration; hence, the ride comfort would be improved which will be shown later in time domain analysis.

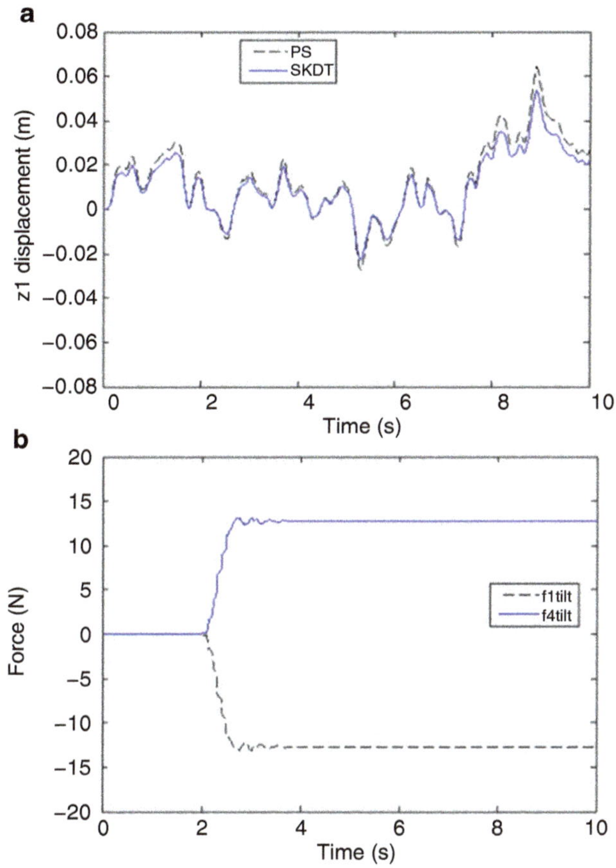

Fig. 6.30 The vehicle sprung mass m_1's vertical displacement for driving scenario two: (**a**) full trajectory and (**b**) short time span

6.4.2 Simulations of Ride Comfort in the Time Domain

In this section, vehicle reaction to the driving scenario using the PS and the SKDT semi-active suspension system is described. Figure 6.53a shows the response of the vehicle's desired tilting angle (faiD) for the steering input signal (delta) and the road bank angle (β) according to the designed direct tilt control (Sect. 5.3.1). The required actuator force to tilt the vehicle according to the desired tilting angle is depicted in Fig. 6.53b.

The time domain responses of the vehicle body vertical, pitch angular, roll angular, lateral acceleration and the vertical displacement of the vehicle front left sprung mass m_1 for this driving scenario are shown in Figs. 6.54, 6.55, 6.56, 6.57, and 6.58, respectively, and it is quite apparent that the displacement of sprung

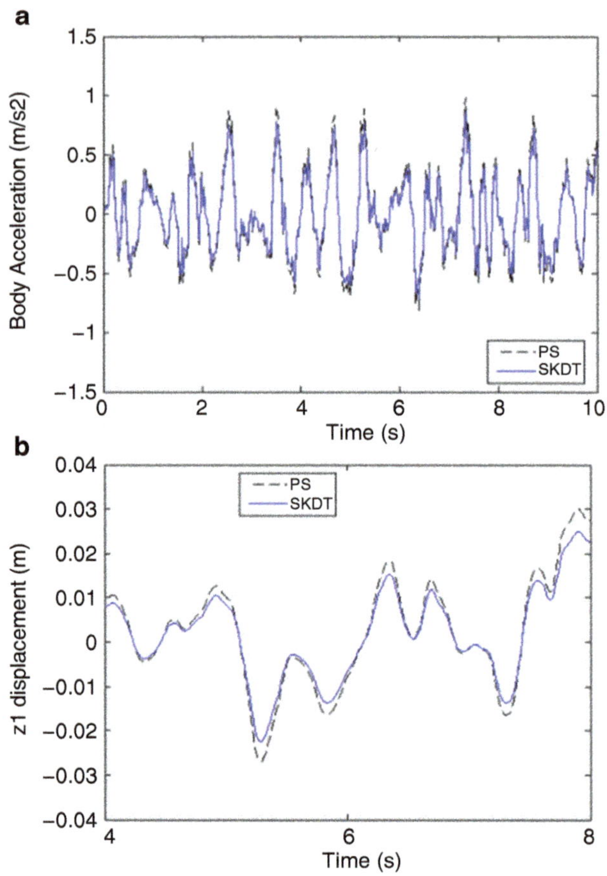

Fig. 6.31 The vehicle body vertical acceleration for driving scenario two: (**a**) full trajectory and (**b**) short time span

mass m_1, the vehicle vertical, pitch angular accelerations and the lateral acceleration of the SK and SKDT systems have decreased compared to the PS system. For better visualization, the responses of the displacement of sprung mass m_1, vehicle vertical, pitch angular roll angular and lateral acceleration and the maximum lateral acceleration value over gravity (a_y/g) have been plotted separately in a short time span.

If the body of the vehicle is parallel with the ground surface, its rollover threshold becomes a function of the vehicle's track width (t) and the centre of gravity height (h) as depicted in Eq. (5.9). The maximum allowable rollover threshold of this vehicle is 1.4578. The rollover threshold defines the maximum lateral acceleration value over gravity (a_y/g) that the vehicle would be able to reach before tipping over. From Fig. 6.59a, b, it is clear that (a_y/g) never exceeds the maximum allowable rollover threshold of this vehicle. Hence, the vehicle remains stable throughout the trajectory of this driving scenario.

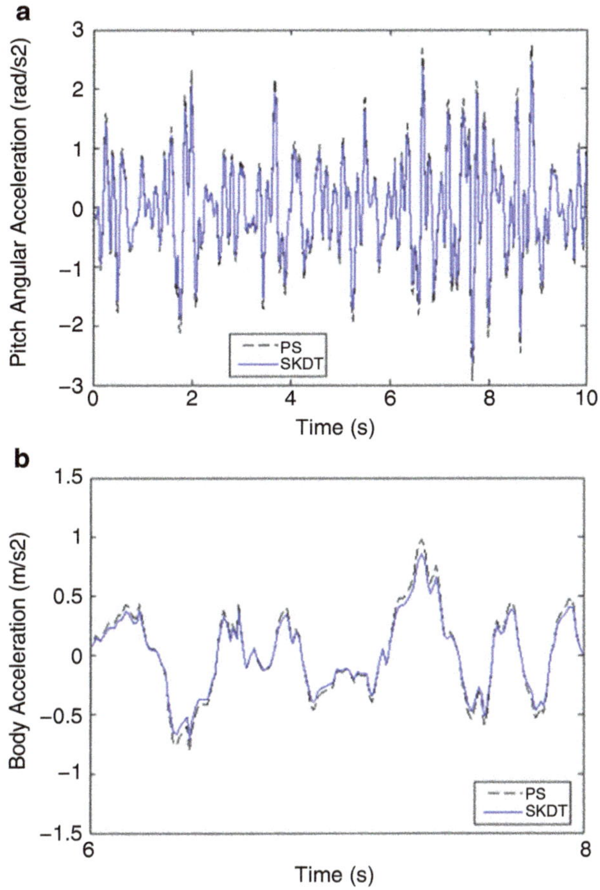

Fig. 6.32 The pitch angular acceleration for driving scenario two: (**a**) full trajectory and (**b**) short time span

Compared to the passive suspension system, the normalized RMS body vertical acceleration (NV) of the SK and SKDT suspension systems has decreased by 10.68% and 12.99%, respectively, as shown in Fig. 6.60. The normalized RMS pitch angular accelerations (NP) of the SK and SKDT suspension systems have decreased by 9.29% and 10.99%, respectively (Fig. 6.61).

However, Fig. 6.62 shows that the normalized RMS roll angular accelerations (NR) of the SK and SKDT systems have increased by 0.61% and 9.99%. Fortunately, the vehicle's roll angular acceleration only accounts for less than 20% of vehicle ride comfort, in terms of ride comfort index specified by the ISO 2631-1 standard. Hence, the vehicle ride comfort would not be affected significantly by the large roll angular acceleration. Moreover, the passengers would feel less lateral acceleration; hence, the ride comfort would be improved. Compared to the PS system, the NL (normalized lateral acceleration) of the SK and

Fig. 6.33 The roll angular acceleration for driving scenario two: (**a**) full trajectory and (**b**) short time span

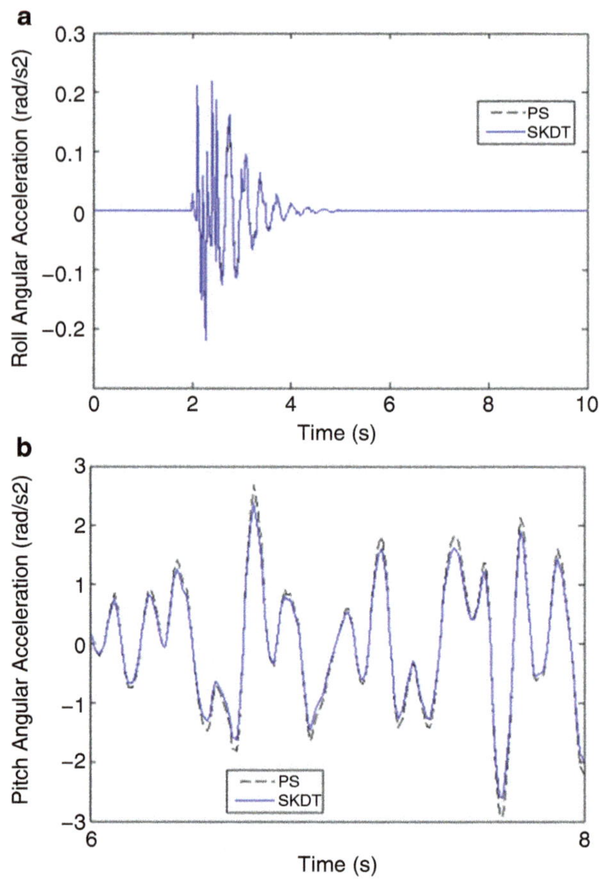

SKDT suspension systems has decreased by 0.46% and 10%, respectively (Fig. 6.63).

As shown in Fig. 6.64, the road handling performances of the SK and SKDT systems have decreased by only 2.09% and 2% compared to the passive suspension system. But the weighted RMS acceleration value of the sprung mass obtained in this simulation is 0.65 and 0.72 for the SK and SKDT systems, respectively, which are within the acceptable range of international code ISO 2631 [143]. Hence, the SK and SKDT suspension systems are able to improve ride comfort of the vehicle effectively while the vehicle is moving on road class C.

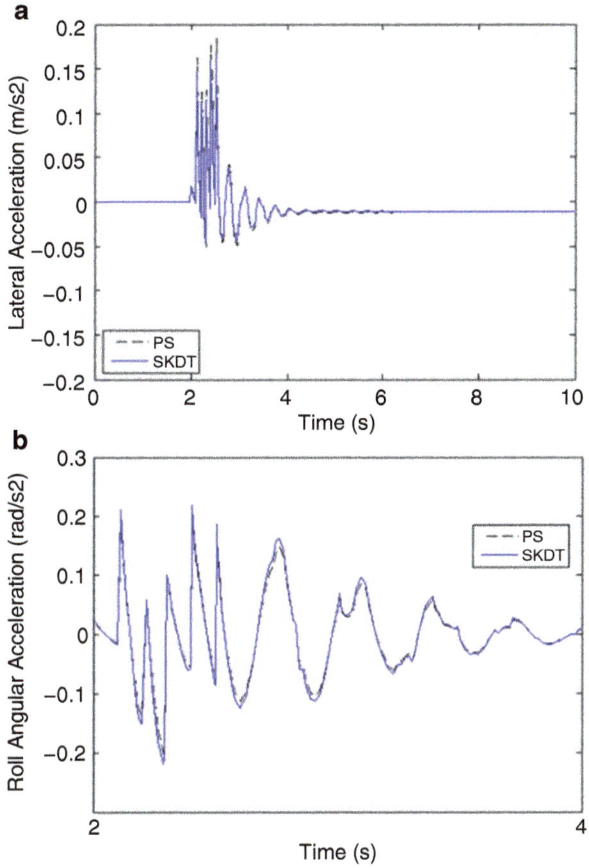

Fig. 6.34 The lateral acceleration for driving scenario two: (**a**) full trajectory and (**b**) short time span

6.5 Conclusions

This chapter describes the conclusion reached from the analysis of the simulations that were carried out to investigate the proposed skyhook control suspension system and the designed direct tilt control system. The SK and SKDT systems were applied to the dynamic model of a full car which considers the road bank angle. At first, the responses of the dynamic model of the full car on different road classes A, B and C defined by ISO 8608 and also combined road profile described in Sect. 4.2 were observed. In this section, PS and SK systems were analysed separately. The behaviours of the vehicle body vertical acceleration and pitch angular acceleration were examined in the frequency domain. While travelling on all the road profiles, it has been ascertained that the vehicle body vertical acceleration and pitch angular acceleration of the SK system remain closer to 0 db than the PS system at a

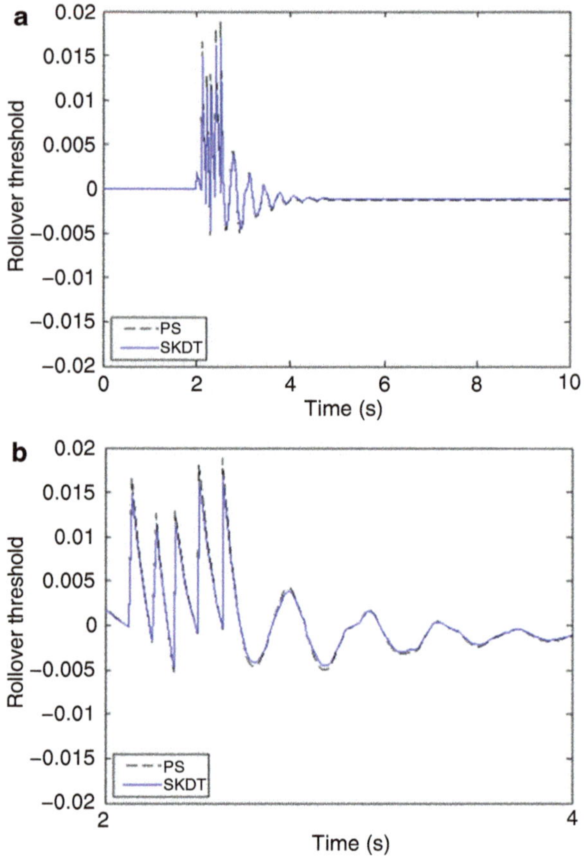

Fig. 6.35 The rollover threshold in driving scenario two: (**a**) full trajectory and (**b**) short time span

wide range of frequencies, particularly in the region where humans are more sensitive to vibrations.

The time domain analysis was also carried out on the full car model using the PS and the SK systems separately. In this subsection, vertical displacement of the vehicle front left sprung mass m_1, vehicle body vertical and pitch angular acceleration were assessed. NV (normalized body vertical acceleration) and NP (normalized pitch angular acceleration) were calculated, and the result is that the SK system significantly reduced the vehicle body vertical acceleration and pitch angular acceleration in all instances compared to the PS system. However, the road handling performance of the SK controlled suspension system decreased slightly compared to the passive suspension system. But the SK system's weighted RMS acceleration value of the sprung mass obtained in this simulation was always within the acceptable range of international code ISO 2631 [143]. Since the vehicle was traversing on a straight road, the differences in roll angular and lateral acceleration of both systems were negligible in this analysis.

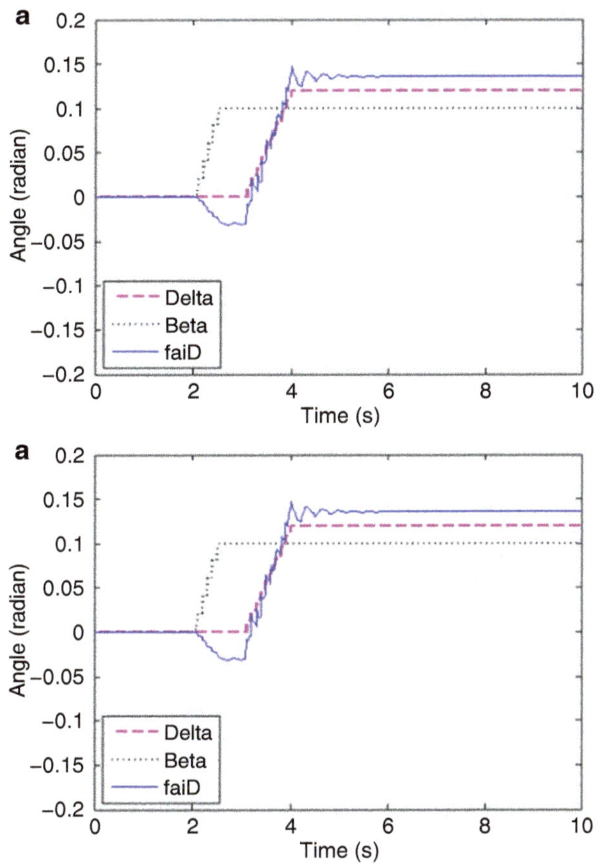

Fig. 6.36 The response of steering and bank angle in driving scenario three: (**a**) desired tilting angle, (**b**) required actuator force

Subsequently, the simulations of the dynamic model of the full car in different driving scenarios (described in Sect. 5.3) were carried out using the SKDT system. Vehicle reaction to the driving scenarios of both the PS and the SKDT semi-active suspension systems was described in this section. The response of the vehicle's desired tilting angle (faiD) for the corresponding steering input signal (delta) and the road bank angle (β) have been shown. The required actuator force to tilt the vehicle according to the desired tilting angle was also observed.

Here the time domain responses of the vertical displacement of the vehicle front left sprung mass m_1, vehicle body vertical, pitch angular, roll angular and lateral acceleration for each driving scenario were shown, and it was quite evident that the displacement of sprung mass m_1, the vehicle vertical, pitch angular and lateral acceleration of the SKDT system have decreased compared to the PS system. Compared to the passive suspension system, the NV and NP of the SKDT suspension system have decreased significantly. However, the normalized

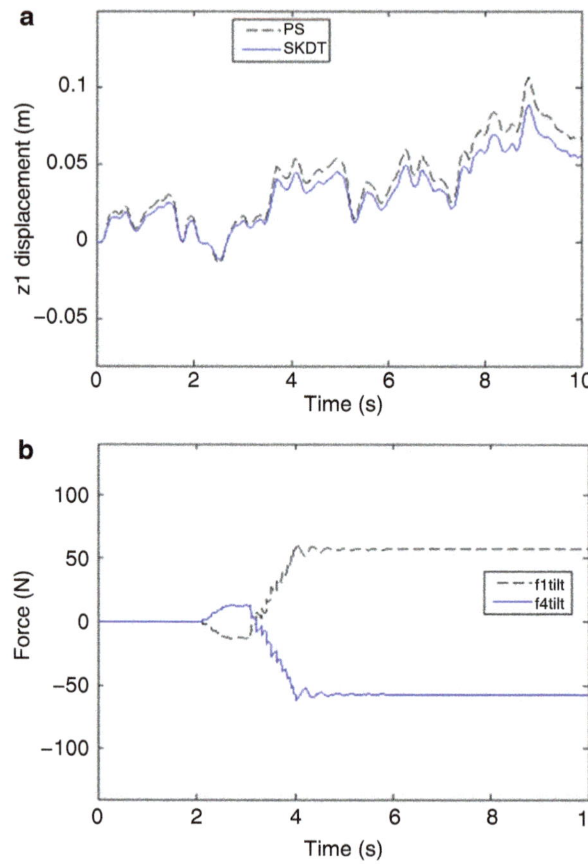

Fig. 6.37 The vehicle sprung mass m_1's vertical displacement for driving scenario three: (**a**) full trajectory and (**b**) short time span

RMS of the roll angular acceleration (NR) of the SKDT system has increased significantly for the tilting action. Fortunately, the vehicle's roll angular acceleration only accounts for less than 20% of vehicle ride comfort, in terms of ride comfort index specified by the ISO 2631-1 standard. Hence, the vehicle ride comfort would not be affected significantly by the large roll angular acceleration. Moreover, the appropriate tilting action reduces the lateral acceleration experienced by passengers and improved the vehicle ride comfort, as the NL (normalized lateral acceleration) of the SKDT suspension system has decreased by 10.01% compared to the PS system. The stability of the resulting closed-loop system was then investigated.

At the end of this chapter, the combined simulation was discussed. Road class C and driving scenario four were used to evaluate the performance of the PS, SK and SKDT systems. It has been demonstrated that the proposed skyhook control system on its own can improve the vehicle ride comfort keeping the road handling performance within an acceptable range. The SKDT system shows that this control

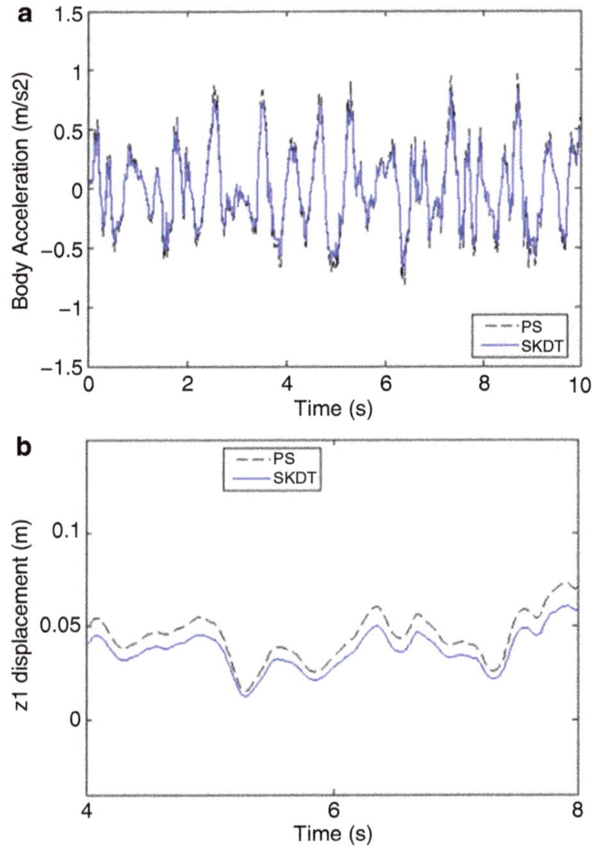

Fig. 6.38 The vehicle body vertical acceleration for driving scenario three: (**a**) full trajectory and (**b**) short time span

strategy is capable of tilting a vehicle inward which would act against the lateral acceleration resulting from steering manoeuvres. The designed SKDT system was capable of maintaining the system's tilt state by keeping the vehicle stable and safe from rollover. This manuscript has indicated the potential of the SKDT suspension system for improving cornering performances of vehicles and paves the way for future work on an integrated system for chassis control.

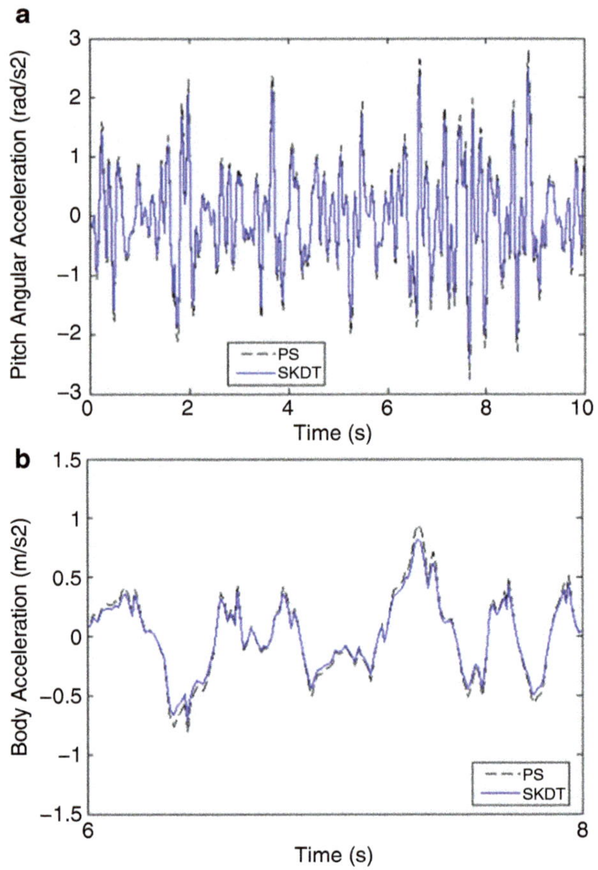

Fig. 6.39 The pitch angular acceleration for driving scenario three: (**a**) full trajectory and (**b**) short time span

Fig. 6.40 The roll angular acceleration for driving scenario three: (a) full trajectory and (b) short time span

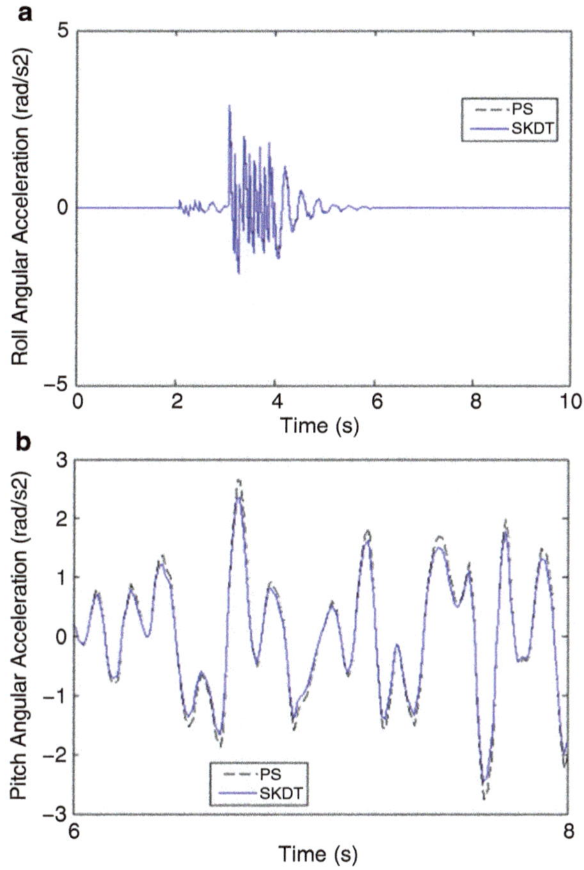

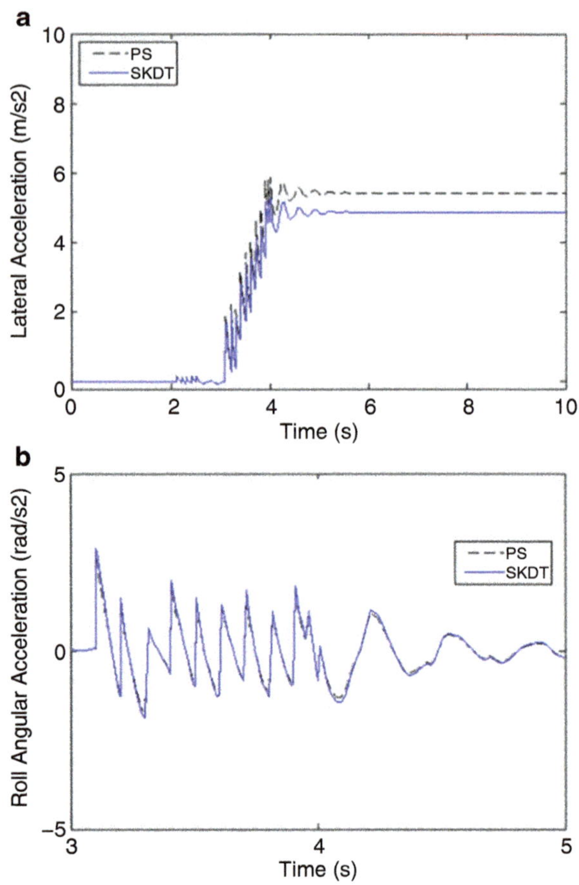

Fig. 6.41 The lateral acceleration for driving scenario three: (**a**) full trajectory and (**b**) short time span

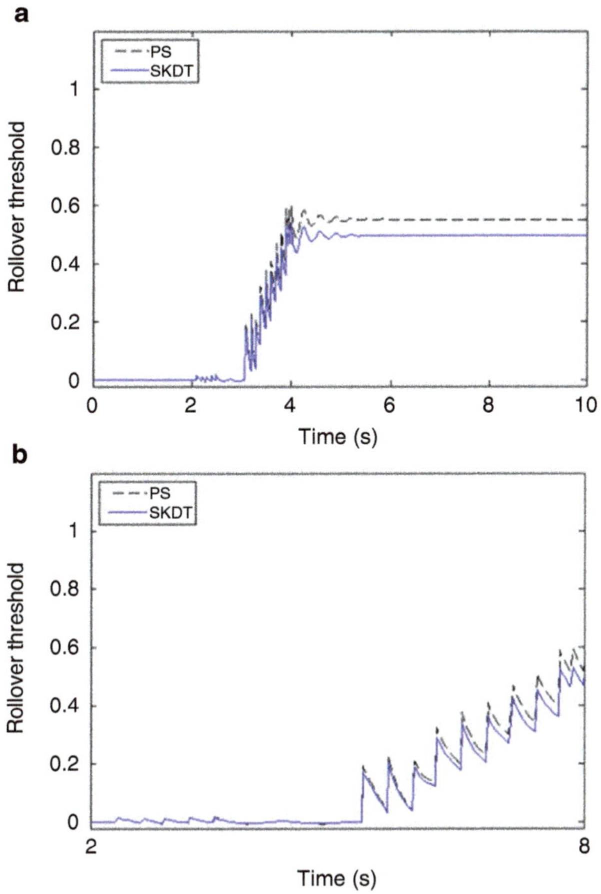

Fig. 6.42 The rollover threshold in driving scenario three: (**a**) full trajectory and (**b**) short time span

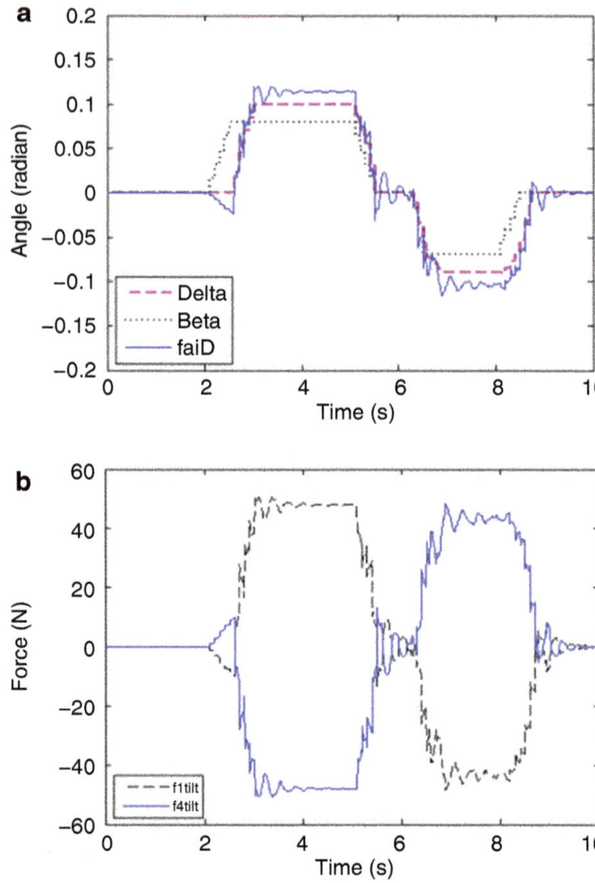

Fig. 6.43 The response of steering and bank angle in driving scenario four: (**a**) desired tilting angle, (**b**) required actuator force

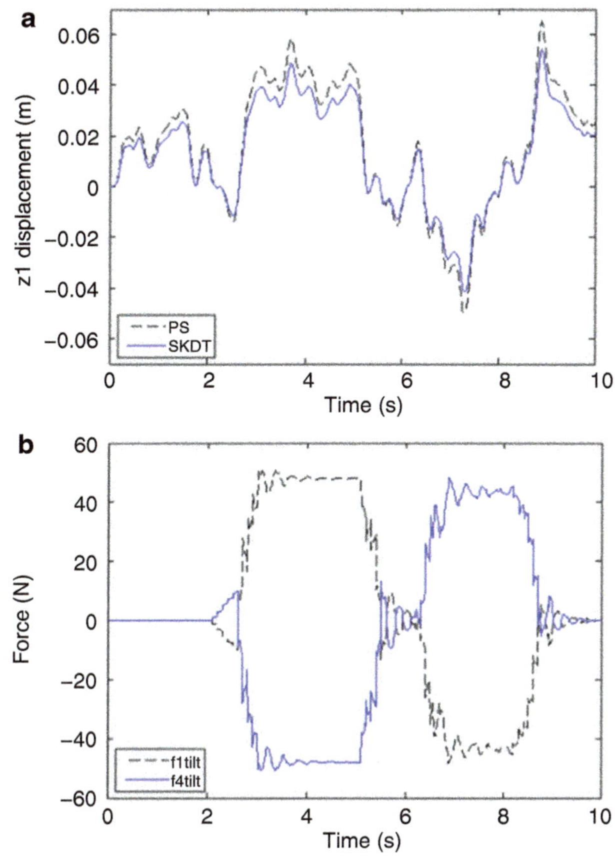

Fig. 6.44 The vehicle sprung mass m_1's vertical displacement for driving scenario four: (**a**) full trajectory and (**b**) short time span

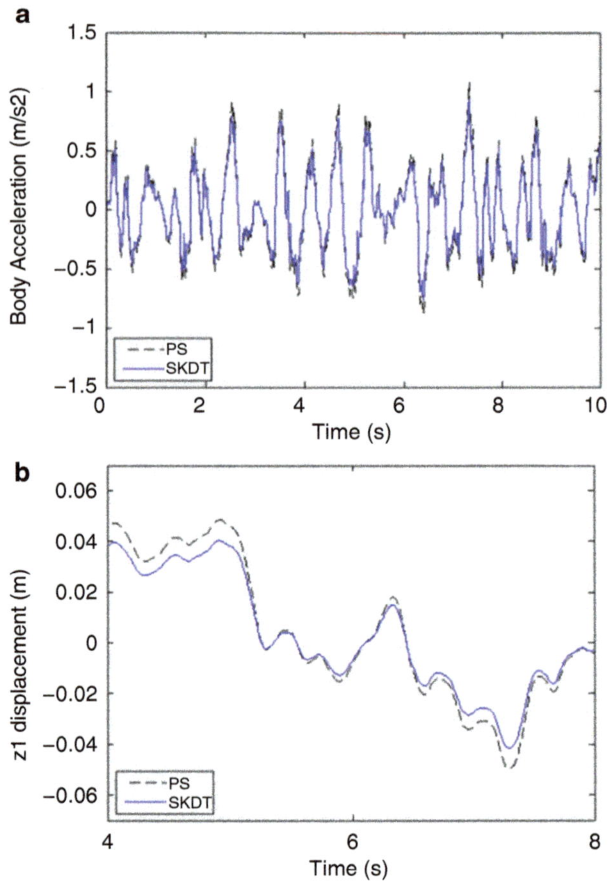

Fig. 6.45 The vehicle body vertical acceleration for driving scenario four: (**a**) full trajectory and (**b**) short time span

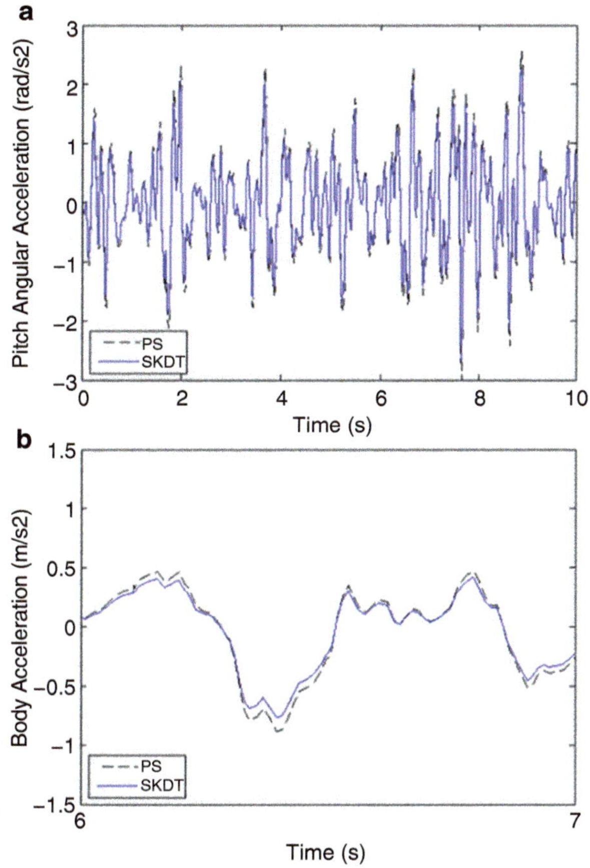

Fig. 6.46 The pitch angular acceleration for driving scenario four: (**a**) full trajectory and (**b**) short time span

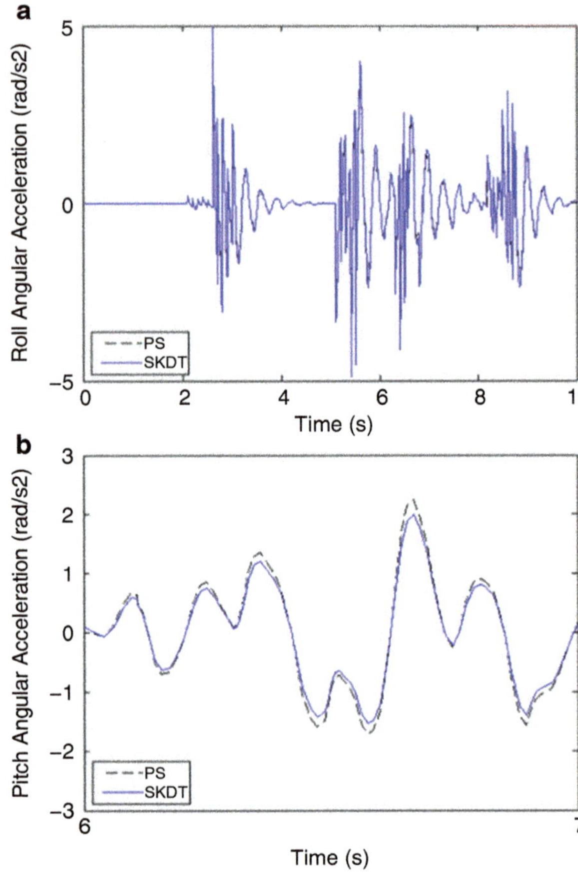

Fig. 6.47 The roll angular acceleration for driving scenario four: (**a**) full trajectory and (**b**) short time span

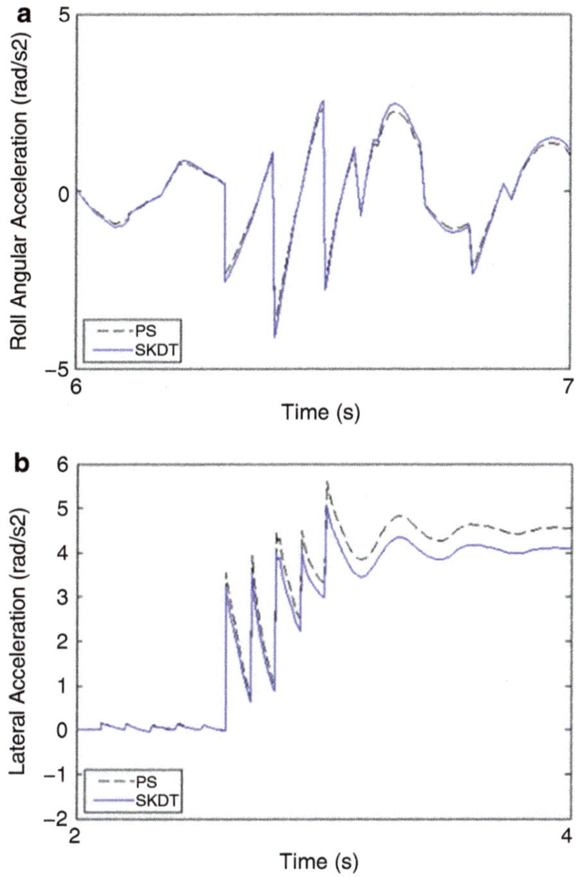

Fig. 6.48 The lateral acceleration for driving scenario four: (**a**) full trajectory and (**b**) short time span

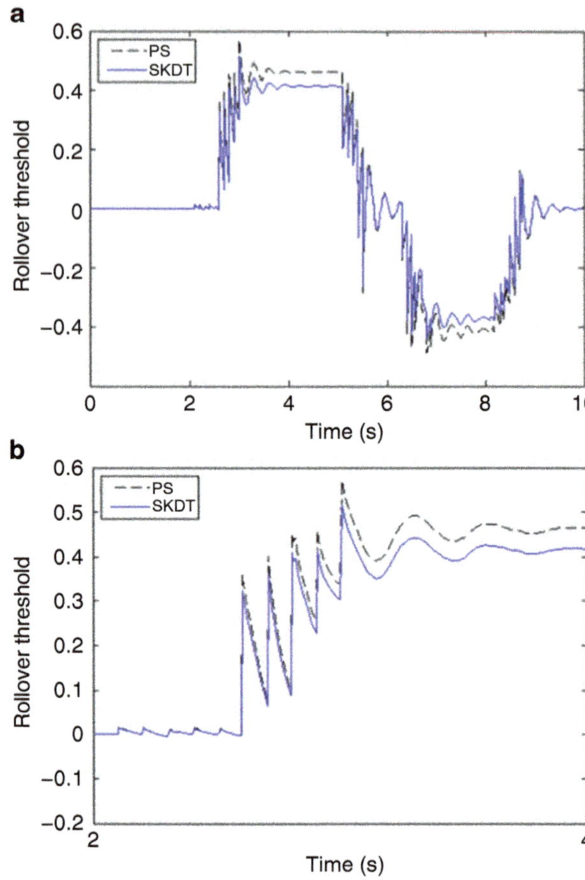

Fig. 6.49 The rollover threshold in driving scenario four: (**a**) full trajectory and (**b**) short time span

Fig. 6.50 The frequency domain response of the car body vertical acceleration: (**a**) at narrow frequency range and (**b**) at broad frequency range

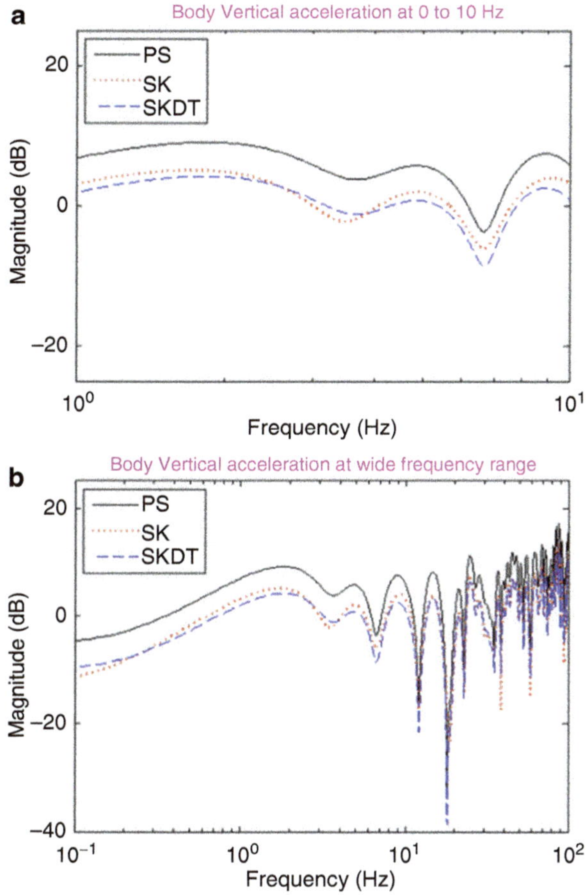

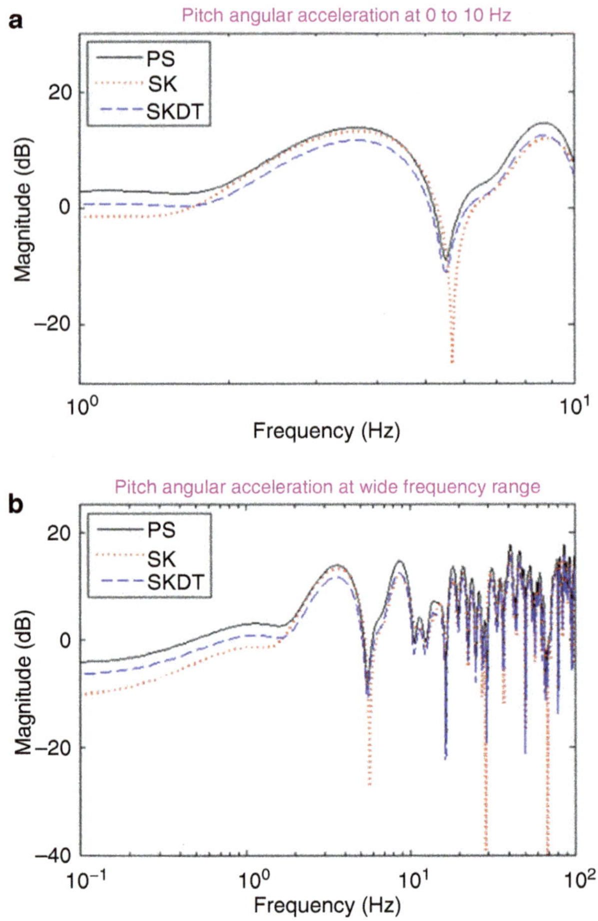

Fig. 6.51 The frequency domain response of the car body pitch angular acceleration (**a**) at narrow frequency range and (**b**) at broad frequency range

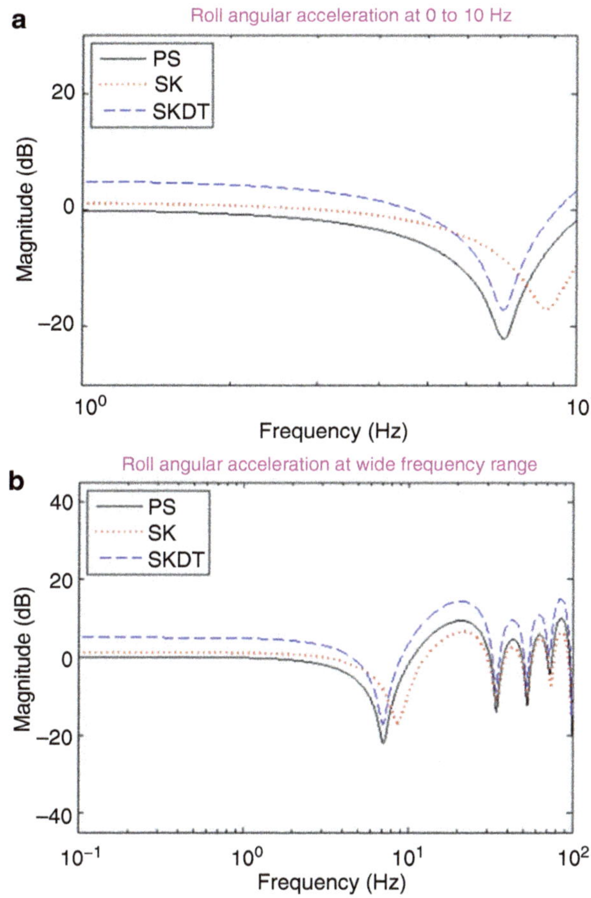

Fig. 6.52 The frequency domain response of the car body roll angular acceleration (**a**) at narrow frequency range and (**b**) at broad frequency range

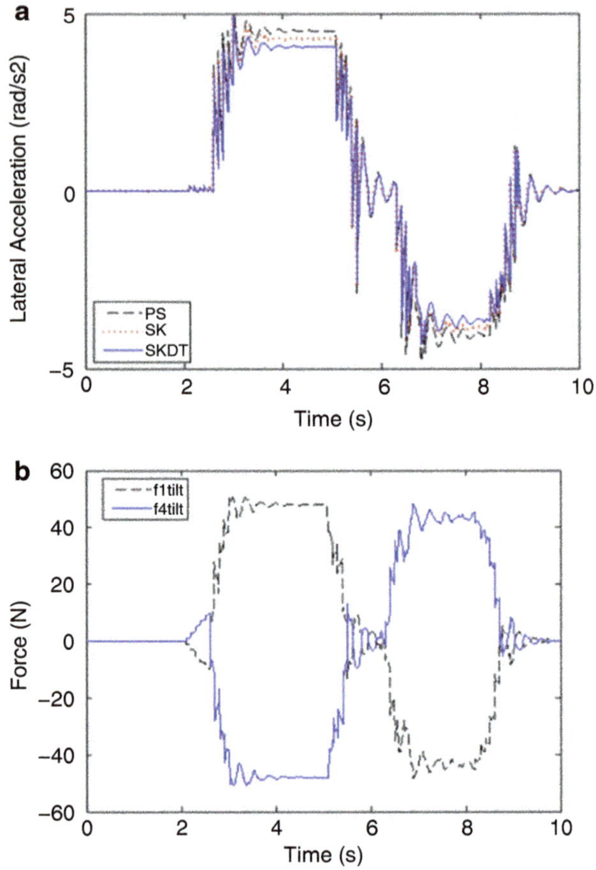

Fig. 6.53 The response of steering and bank angle in driving scenario four and road class C: (**a**) desired tilting angle, (**b**) required actuator force

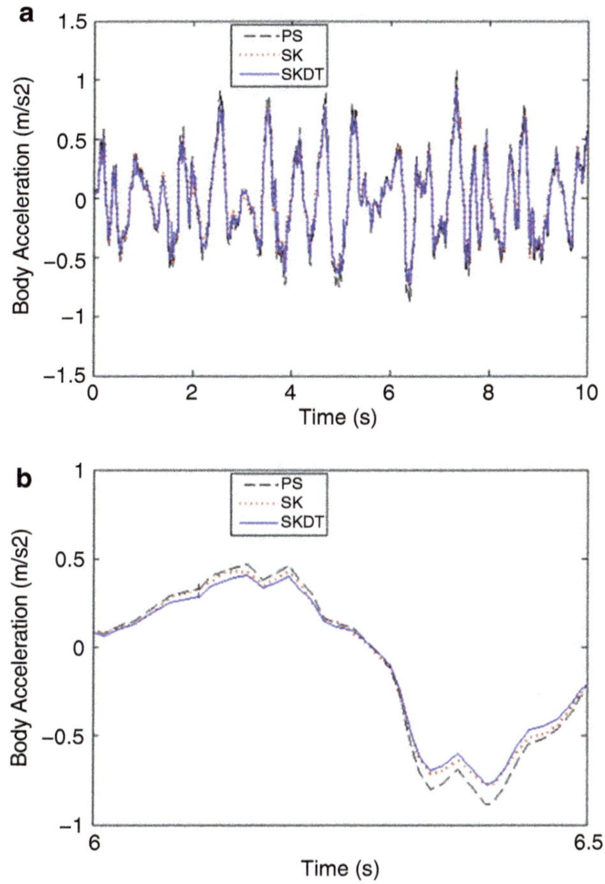

Fig. 6.54 The vehicle body vertical acceleration for driving scenario four and road class C: (**a**) full trajectory and (**b**) short time span

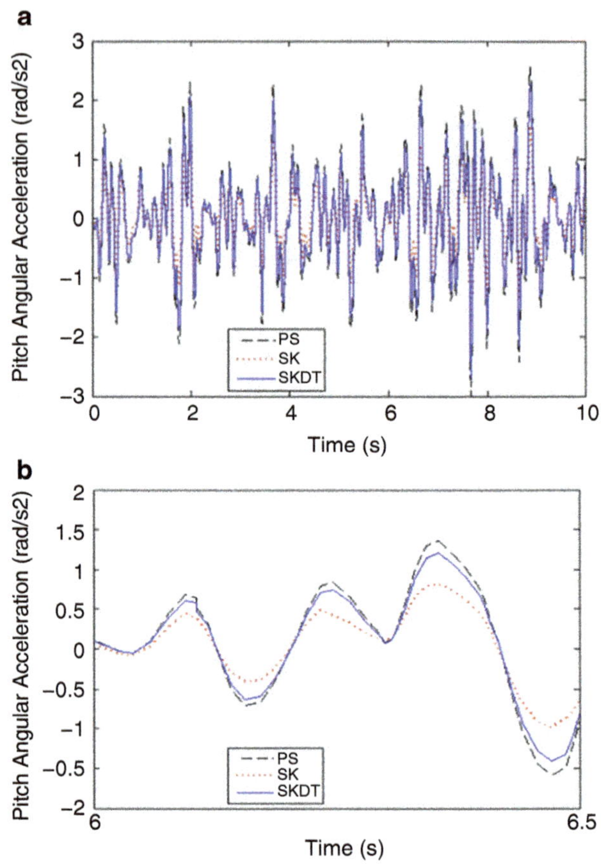

Fig. 6.55 The pitch angular acceleration for driving scenario four and road class C: (**a**) full trajectory and (**b**) short time span

Fig. 6.56 The roll angular acceleration for driving scenario four and road class C: (**a**) full trajectory and (**b**) short time span

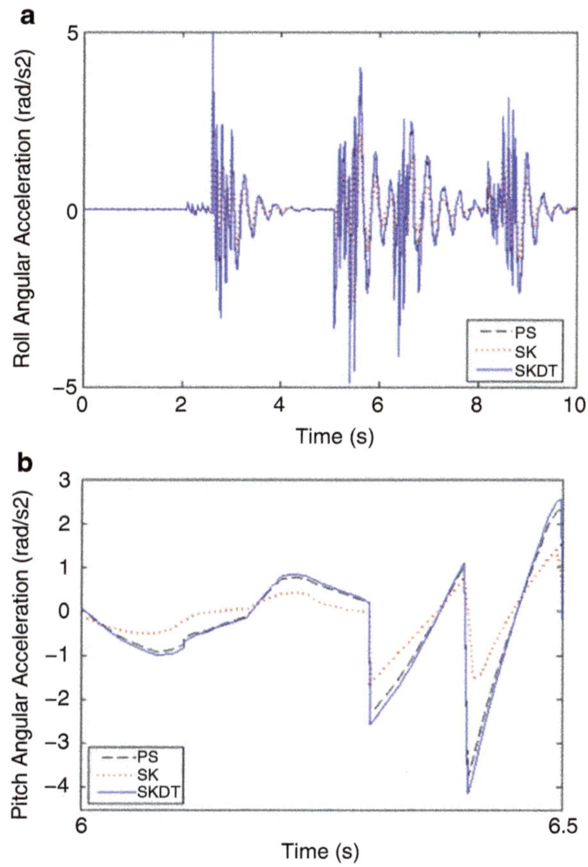

Fig. 6.57 The lateral acceleration for driving scenario four and road class C: (a) full trajectory and (b) short time span

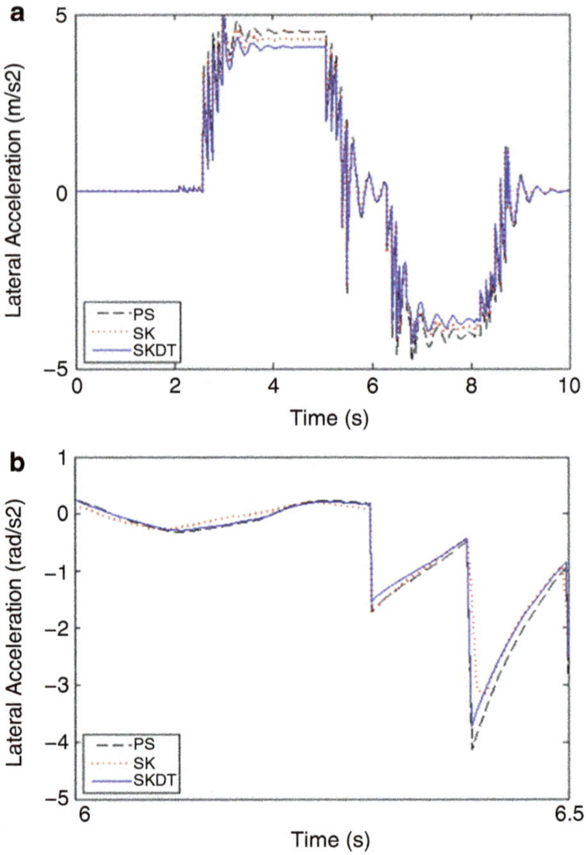

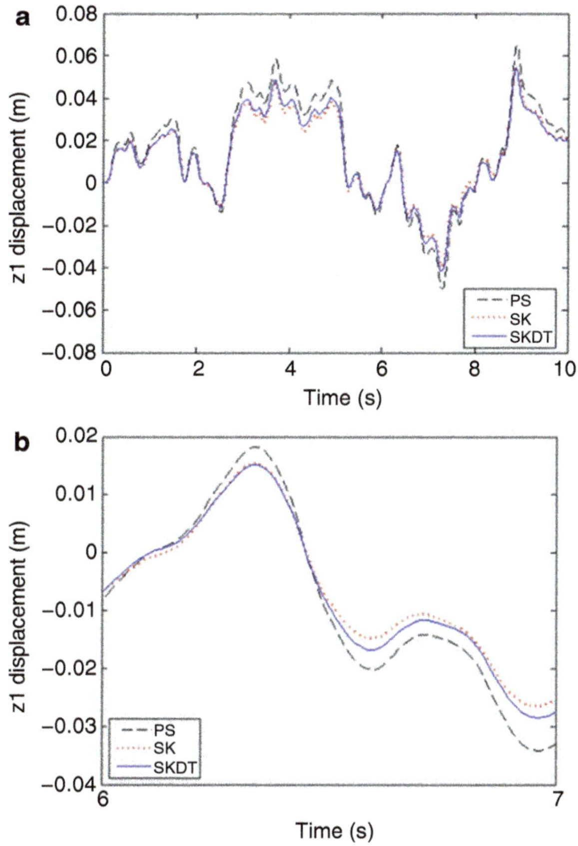

Fig. 6.58 The vehicle sprung mass m_1's vertical displacement for driving scenario four and road class C: (**a**) full trajectory and (**b**) short time span

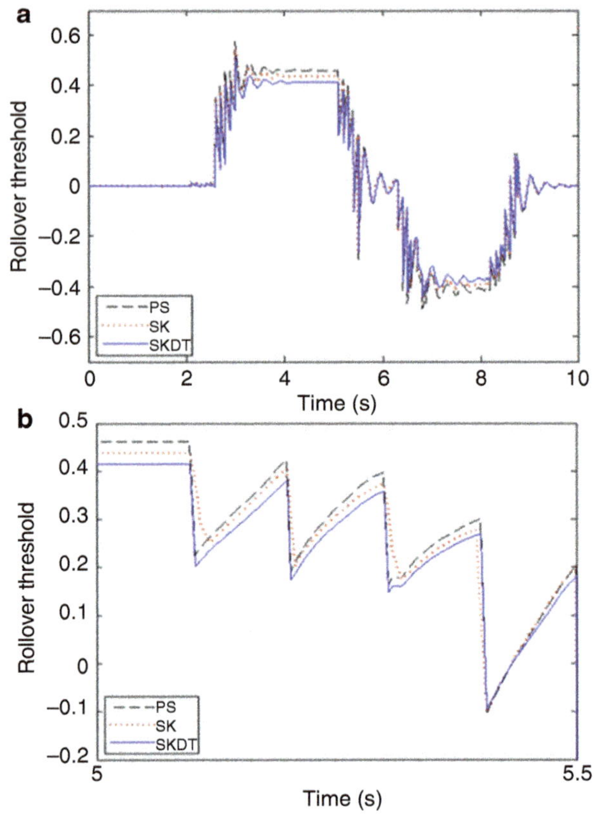

Fig. 6.59 The rollover threshold in driving scenario four and road class C: (**a**) full trajectory and (**b**) short time span

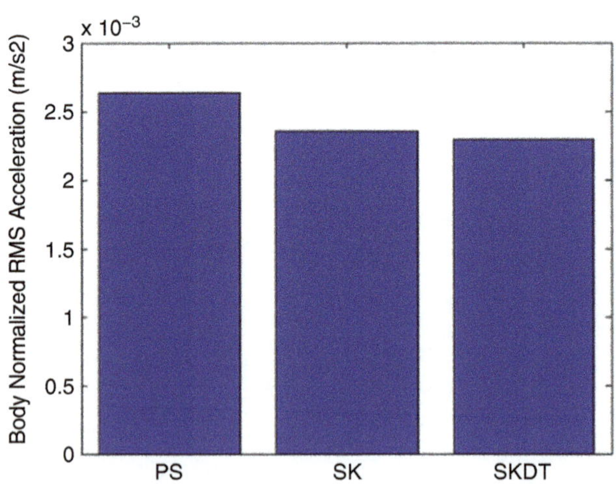

Fig. 6.60 Vehicle body vertical acceleration comparison

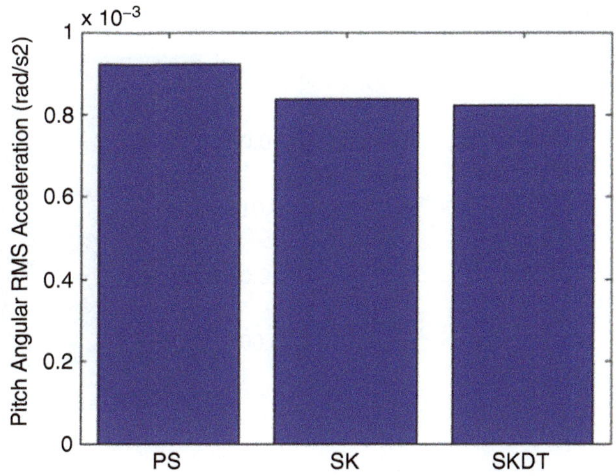

Fig. 6.61 Vehicle body pitch angular acceleration comparison

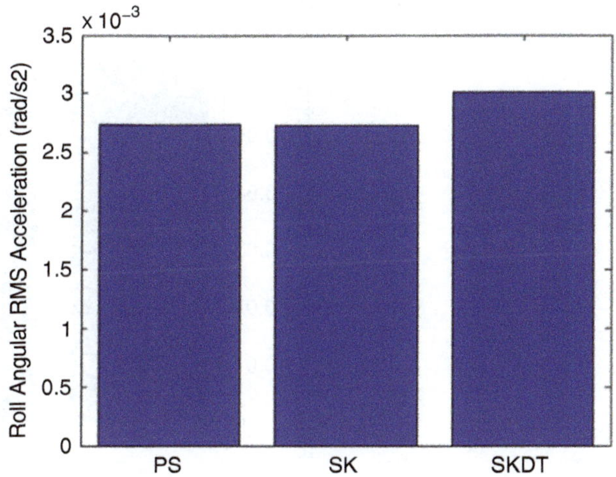

Fig. 6.62 Vehicle body roll angular acceleration comparison

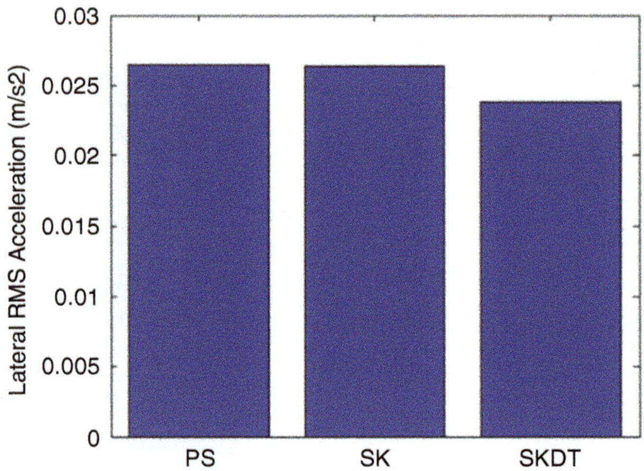

Fig. 6.63 Vehicle body lateral acceleration comparison

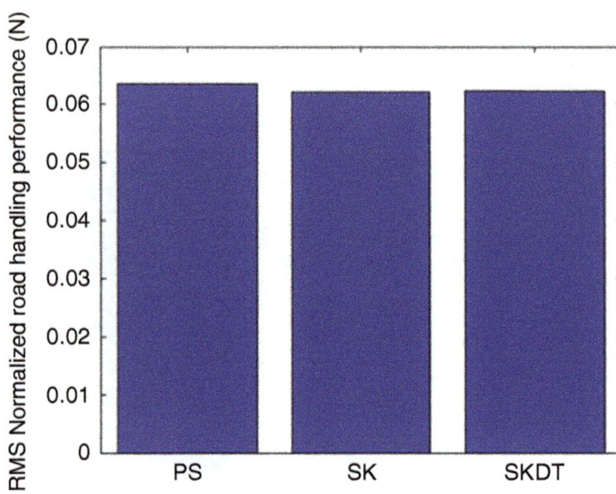

Fig. 6.64 Vehicle road handling performance comparison

Chapter 7
Experimental Analysis of Full Car Model

Abstract In this chapter, the analysis of the dynamics of a full car model is presented. It incorporates the response of the Quanser quarter-car suspension plant as one of the four wheels of the full car model. Section 7.1 describes the environment of the experimental analysis and the parameters of the full car that emulated from the Quanser quarter-car suspension plant. The vehicle performance analysis using the Quanser quarter-car suspension plant at the front left and rear right wheel of the full car model is presented separately in Sects. 7.2 and 7.3. The performance of the combined approach where the proposed skyhook controller is activated along with the direct tilt control is evaluated in Sects. 7.2 and 7.3 at frequency domain and time domain.

7.1 Experimental Environment

One objective of the suspension control system design in this manuscript is to tilt the vehicle inward when it is steered. The proposed skyhook control and the tilting mode of the system were investigated using the Quanser suspension plant. The real-time simulation environment with the experimental model was set up in such a way that the road profile input interacting with the front left wheel of the full vehicle model was fed into the Quanser quarter-car suspension plant using the Simulink model illustrated in Fig. 7.1. The output of the Quanser suspension plant was then fed back to the full car model to complete the real-time simulation. An analysis was carried out using the Quanser quarter-car suspension plant as the rear right wheel of the full car model. The Quanser suspension system has three masses, each supported by two springs as shown in Fig. 4.8. The vehicle body is represented by the upper mass (blue) supported above the suspension while the middle mass (red) represents one of the vehicle's wheels. A programmable motor is used to actuate the upper mass. The lower plate (silver) emulates the road surface by moving vertically. The performance criteria can be formulated into a mathematical model. The ride comfort is measurable through an accelerometer located on the

S. Kashem et al., *Vehicle Suspension Systems and Electromagnetic Dampers*,
Springer Tracts in Mechanical Engineering, DOI 10.1007/978-981-10-5478-5_7

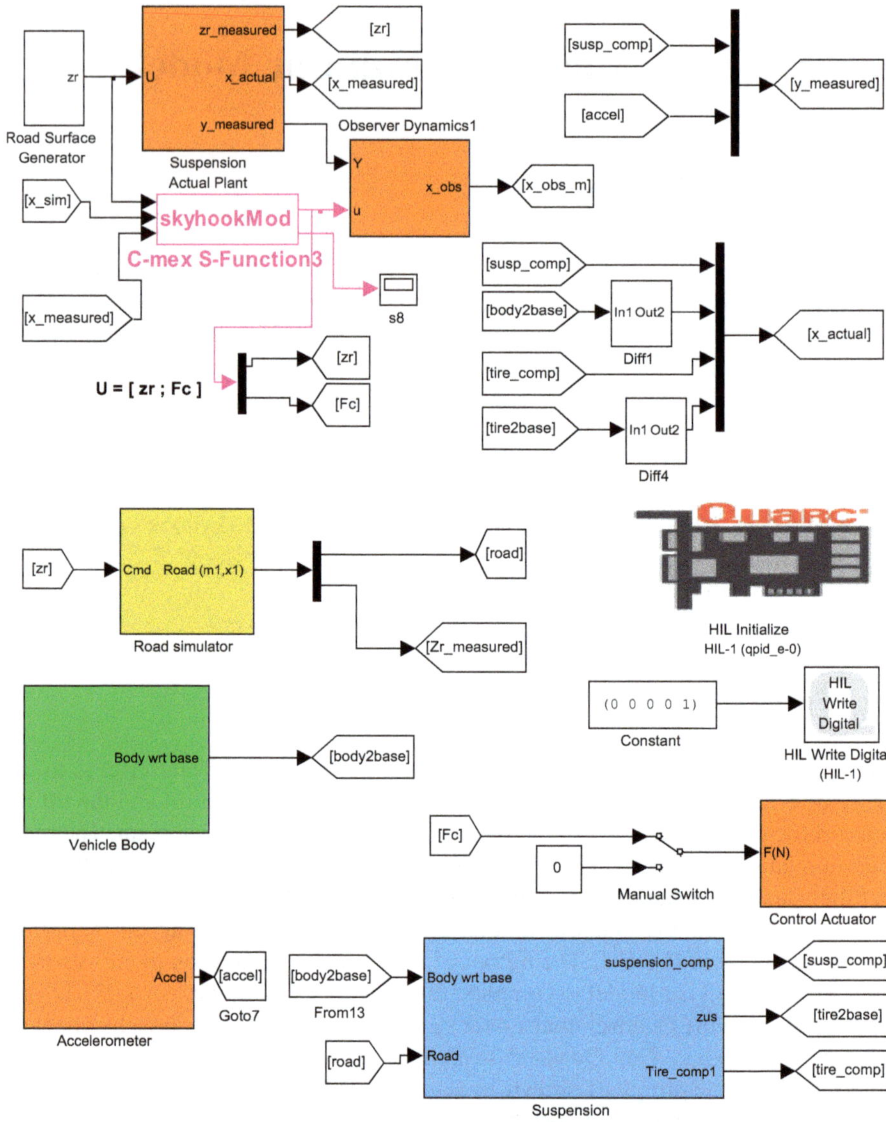

Fig. 7.1 Quanser simulink model

sprung mass. Suspension travel is measured using a linear capstan mechanism. The parameters that have been used in this analysis are listed in Table 7.1.

The objective of this section is to describe the evaluation of the SK system (the proposed skyhook suspension system) and the SKDT system (the direct tilt control is activated along with the proposed skyhook controller). The performance of the SK and SKDT systems are compared to the PS system (the passive suspension

Table 7.1 Nominal parameter values used in the experiment

Parameter	Definition	Values
m_i ($i = 1, 2, 3, 4$)	Vehicle unsprung masses	1 (kg)
m	Sprung mass	10 (kg)
M_t	Total mass of the vehicle	14 (kg)
k_{i1} ($i = 1, 2, 3, 4$)	Tyre spring coefficient	2500 (N/m)
K_{i2} ($i = 1, 2, 3, 4$)	Suspension spring coefficient	900 (N/m)
$c_{t1} \sim c_{t4}$	Tyre damping coefficient	5 (N-s/m)
$c_1 \sim c_4$	Suspension damping coefficient	7.5 (N-s/m)
a	Distance from CG to front wheel	0.732 (m)
b	Distance from CG to rear wheel	0.982 (m)
c, d	Half of wheel base	0.452 (m)
h	Height of CG of the vehicle	0.35 (m)
C_{af}, C_{ar}	Cornering stiffness of the tyre	16,339 (N/rad)
I_x	Roll moment of inertial	1.13 (kg m^2)
I_y	Pitch moment of inertial	4.79 (kg m^2)
I_z	Yaw moment of inertial	5.34 (kg m^2)
g	Gravity acceleration	9.8 (m/s^2)

system). In this analysis, the vehicle experienced road class C defined by ISO 8608 described in Sect. 4.2. The vehicle trajectory is defined by driving scenario four (Sect. 5.4). This driving scenario has been used as the vehicle experiences both left and right turns. The vehicle performance analysis using the Quanser quarter-car suspension plant at the front left and rear right wheel of the full car model is presented separately in the following sections (Fig. 7.2).

7.2 Quanser Plant at Front Left Suspension

In this section, the experimental analysis carried out on the full car suspension system is described. Here the Quanser suspension plant is considered to be in use at the front left suspension of the full car suspension model.

Figure 7.3 shows that there is a slight difference in the vehicle front left sprung mass vertical displacement in the simulation environment and experimental setup. This is because the response time of the experimental and simulation environment is slightly different. But the comparison between the simulation and experimental analysis has not been carried out in this section as the simulation of a full car model similar to a standard production vehicle has been carried out in a previous chapter

Fig. 7.2 Quanser intelligent suspension plant

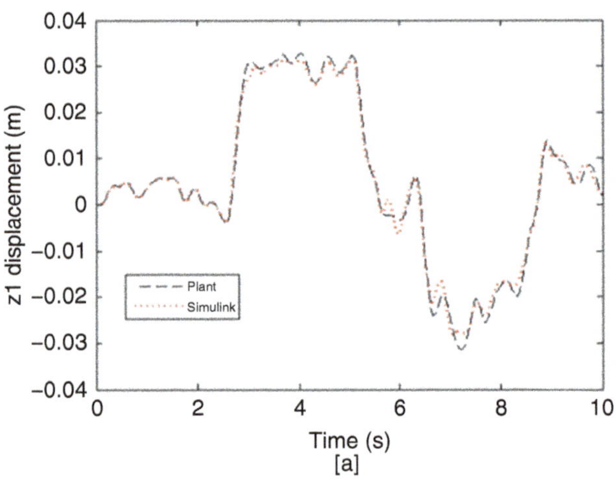

Fig. 7.3 The vehicle front left sprung mass vertical displacement

already. The frequency domain analysis of the vehicle vertical, pitch angular and roll angular acceleration responses to road disturbance class C has been presented in Sect. 7.2.1. The response of the vehicle's desired tilting angle (faiD) for the steering input signal (Delta) and the road bank angle (β) according to the designed direct tilt control system has been described in Sect. 7.2.2. In this subsection, the time domain analysis has also been conducted on the vehicle body vertical, pitch angular, roll

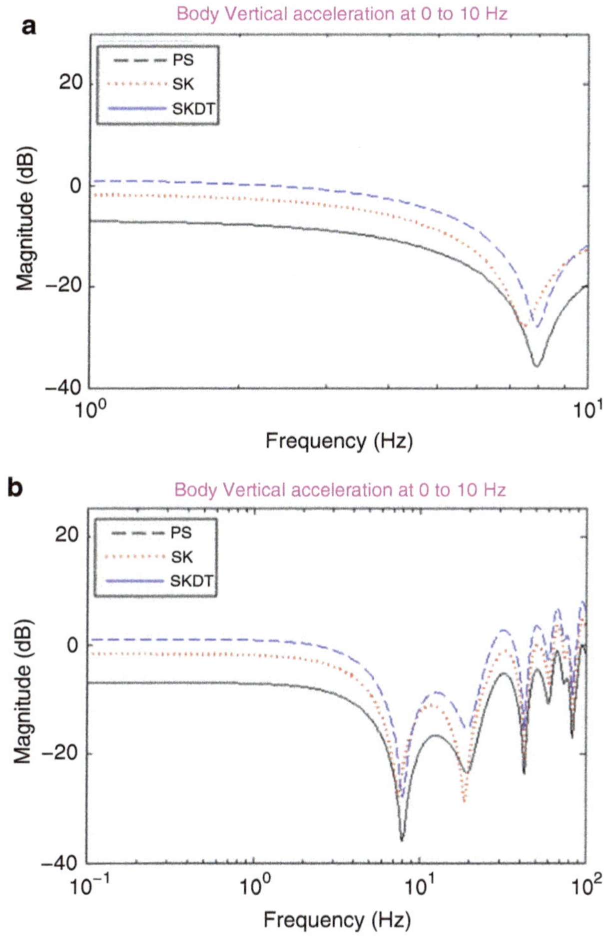

Fig. 7.4 The frequency response of vehicle body vertical acceleration: (**a**) at narrow frequency range and (**b**) at broad frequency range

angular, lateral acceleration and vehicle front left sprung mass vertical displacement.

7.2.1 Experiments of Ride Comfort in the Frequency Domain

Figures 7.4 and 7.5 show the frequency responses of the vehicle vertical, pitch angular and roll angular acceleration in response to road disturbance class C. For better visualization, the responses of vehicle vertical, pitch angular and roll angular accelerations at 0–10 Hz have been plotted separately in a short time span in Figs. 7.4a, 7.5a, and 7.6a, respectively.

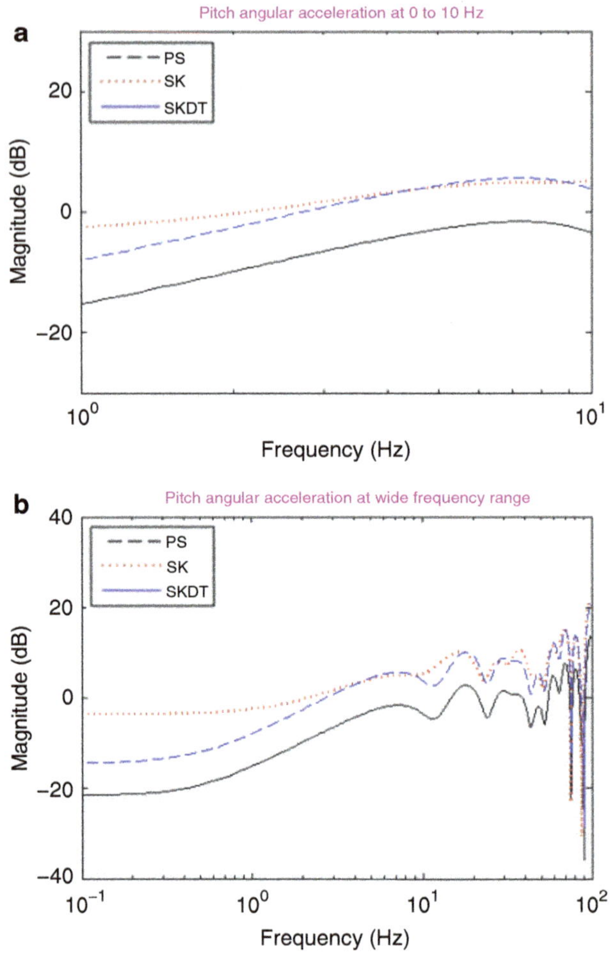

Fig. 7.5 The frequency domain response of the car body pitch angular acceleration: (**a**) at narrow frequency range and (**b**) at broad frequency range

Studies show that human beings are very sensitive to the vertical motion in the frequency range of 4–8 Hz [134]. So while designing a vehicle suspension system, sprung mass acceleration tolerance should be kept to a minimum in the frequency range mentioned above. From Figs. 7.4 and 7.5, it is quite evident that the vehicle body vertical acceleration and pitch angular acceleration of the SK and SKDT systems remained closer to 0 db than with the PS system at a wide range of frequencies, especially at frequencies where humans are more sensitive to vibrations (Figs. 7.4a and 7.5a).

However, the frequency responses of roll angular acceleration of the SK and SKDT systems have deteriorated compared to the PS system as shown in Fig. 7.6. Fortunately, the roll angular acceleration of the vehicle only accounts for less than 20% of vehicle ride comfort, in terms of the ride comfort index specified by ISO 2631-1 standard. Hence the vehicle ride comfort would not be affected significantly

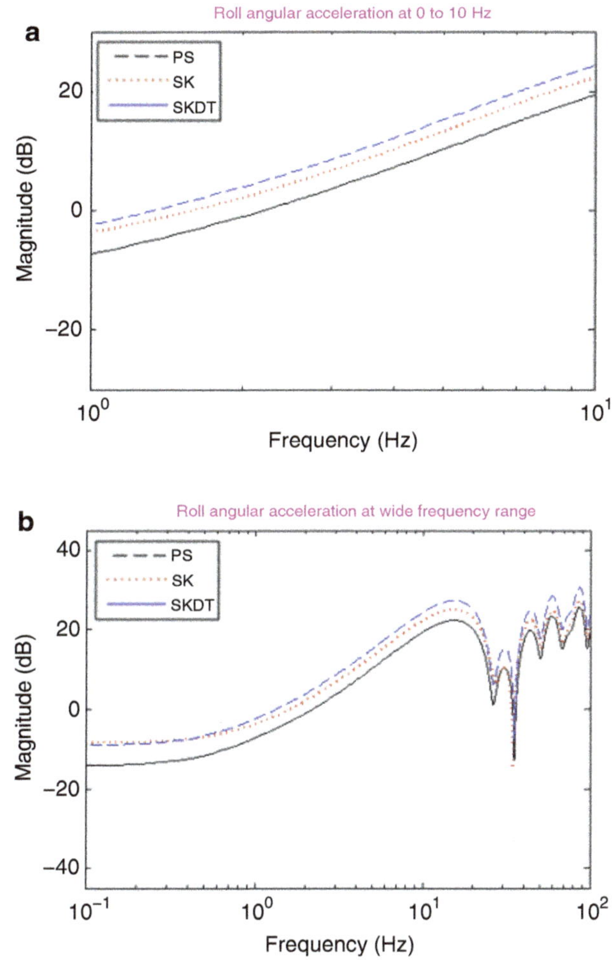

Fig. 7.6 The frequency domain response of the car body roll angular acceleration: (**a**) at narrow frequency range and (**b**) at broad frequency range

by the large roll angular acceleration. Moreover, the passengers would feel less lateral acceleration; hence the ride comfort would be improved. This is shown in a subsequent section in the time domain analysis.

7.2.2 Experiments of Ride Comfort in the Time Domain

In this section, vehicle reactions to the driving scenario using the PS and the SKDT semi-active suspension systems are described. Figure 7.7a shows the response of the vehicle's desired tilting angle (faiD) for the steering input signal (Delta) and the road bank angle (β) according to the designed direct tilt control system. The

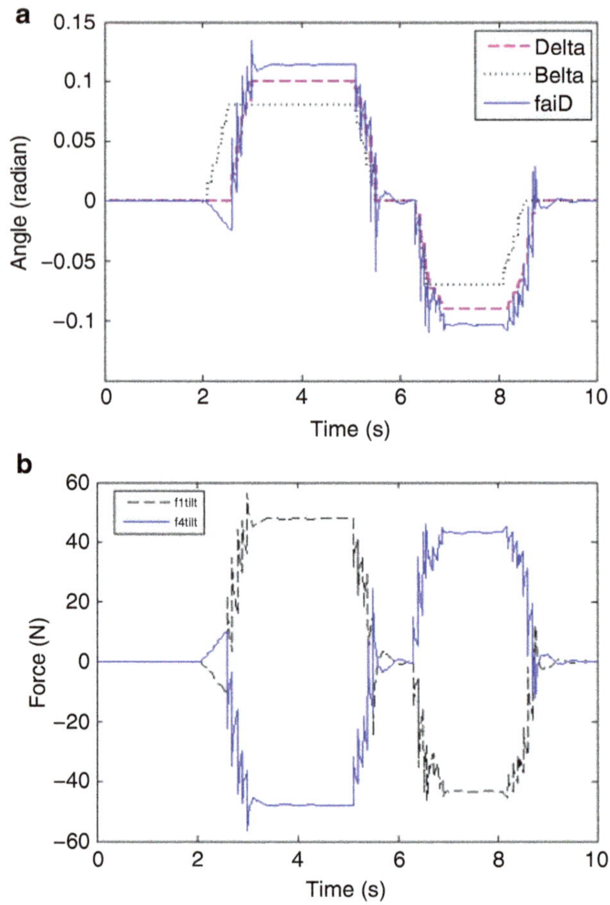

Fig. 7.7 The response of steering and bank angle in driving scenario four and road class C: (**a**) desired tilting angle and (**b**) required actuator force

required actuator force to tilt the vehicle according to the desired tilting angle is depicted in Fig. 7.7b.

The time domain responses of the vehicle body vertical, pitch angular, roll angular, lateral acceleration and the vertical displacement of the vehicle front left sprung mass m_1 for this driving scenario are shown in Figs. 7.8, 7.9, 7.10, 7.11, and 7.12, respectively, and it is quite obvious that the displacement of sprung mass m_1, the vehicle vertical, pitch angular accelerations and the lateral acceleration of the SK and SKDT systems have decreased compared to the PS system. For better visualization, the responses of the displacement of sprung mass m_1, vehicle vertical, pitch angular roll angular and lateral acceleration and the maximum lateral acceleration value over gravity (a_y/g) have been plotted separately in a short time span.

If the body of the vehicle is parallel with the ground surface, its rollover threshold becomes a function of the vehicle's track width (t) and the centre of gravity height (h) as depicted in Eq. (5.9). The maximum allowable rollover

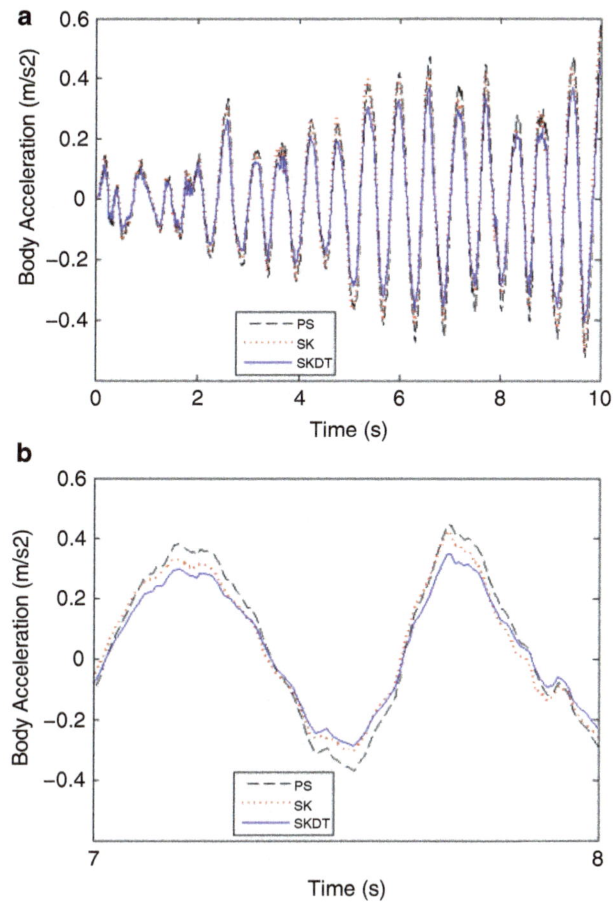

Fig. 7.8 The vehicle body vertical acceleration for driving scenario four and road class C: (**a**) full trajectory and (**b**) short time span

threshold of this vehicle is 1.4578. The rollover threshold defines the maximum lateral acceleration value over gravity (a_y/g) that the vehicle would be able to reach before tipping over. From Fig. 7.13a, b, it is clear that (a_y/g) never exceeds the maximum allowable rollover threshold of this vehicle. Hence the vehicle remains stable throughout the trajectory of this driving scenario.

Compared to the passive suspension system, the normalized RMS body vertical acceleration (NV) of the SK and SKDT suspension systems decreased by 13.88% and 16.96%, respectively, as shown in Fig. 7.14. The normalized RMS pitch angular accelerations (NP) of the SK and SKDT suspension systems decreased by 13.35% and 11%, respectively (Fig. 7.15).

However Fig. 7.16 shows that the normalized RMS roll angular acceleration (NR) of the SK and SKDT systems increased by 4.21% and 10%. Fortunately, the vehicle's roll angular acceleration only accounts for less than 20% of vehicle ride comfort, in terms of the ride comfort index specified by ISO 2631-1 standard. Hence the vehicle ride comfort would not be affected significantly by the large roll

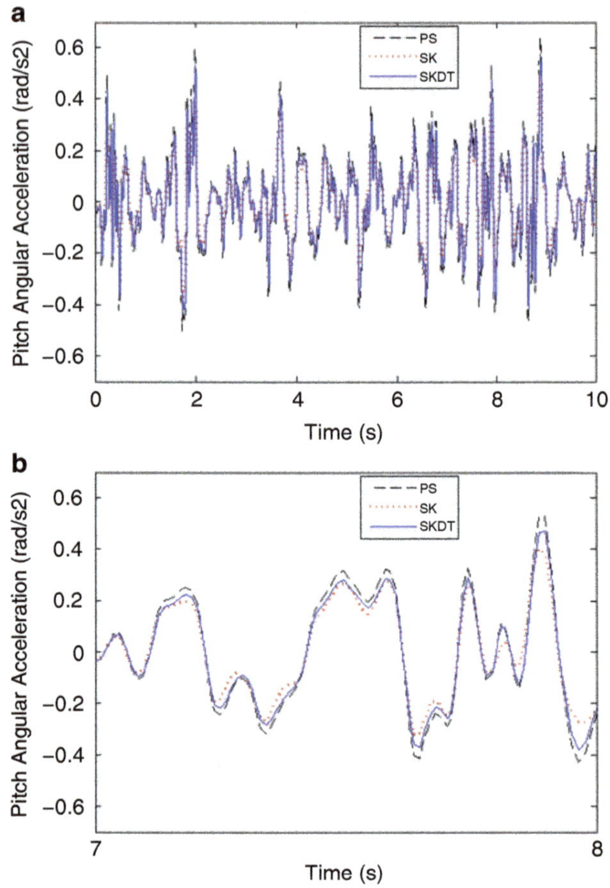

Fig. 7.9 The pitch angular acceleration for driving scenario four and road class C: (**a**) full trajectory and (**b**) short time span

angular acceleration. Moreover, the passengers would feel less lateral acceleration; hence the ride comfort would be improved. Compared to the PS system, the NL (normalized lateral acceleration) of the SK and SKDT suspension systems have decreased by 5.82% and 10.22%, respectively (Fig. 7.17). As shown in Fig. 7.18, the road handling performance of the SK and SKDT systems decreased by only 3.41% and 2.11% compared to the passive suspension system which is within the acceptable range. Hence, the SK and SKDT suspension controllers can effectively improve vehicle ride comfort while the vehicle is moving on road class C.

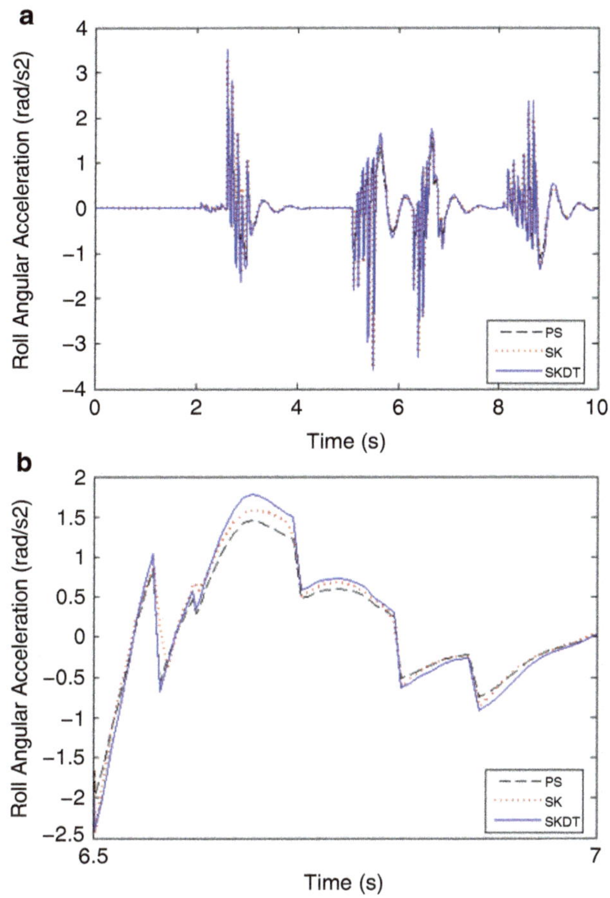

Fig. 7.10 The roll angular acceleration for driving scenario four and road class C: (**a**) full trajectory and (**b**) short time span

7.3 Quanser Plant at Rear Right Suspension

In this section, experimental analysis was carried out on a full car suspension system. Here the Quanser suspension plant is considered to be in use at the rear right suspension of the full car suspension model.

Figure 7.19 shows that there is a slight difference in the vehicle rear right sprung mass vertical displacement with the simulation environment and experimental setup. This is because the response time of the experimental and simulation environments is slightly different. But the comparison between the simulation and experimental analysis is not discussed in this section as the simulation of the full car model similar to a standard production vehicle was discussed in the previous chapter. The frequency domain analysis of the vehicle vertical, pitch angular and roll angular accelerations in response to the road disturbance class C was presented in Sect. 7.3.1. The response of the vehicle's desired tilting angle

Fig. 7.11 The lateral acceleration for driving scenario four and road class C: (**a**) full trajectory and (**b**) short time span

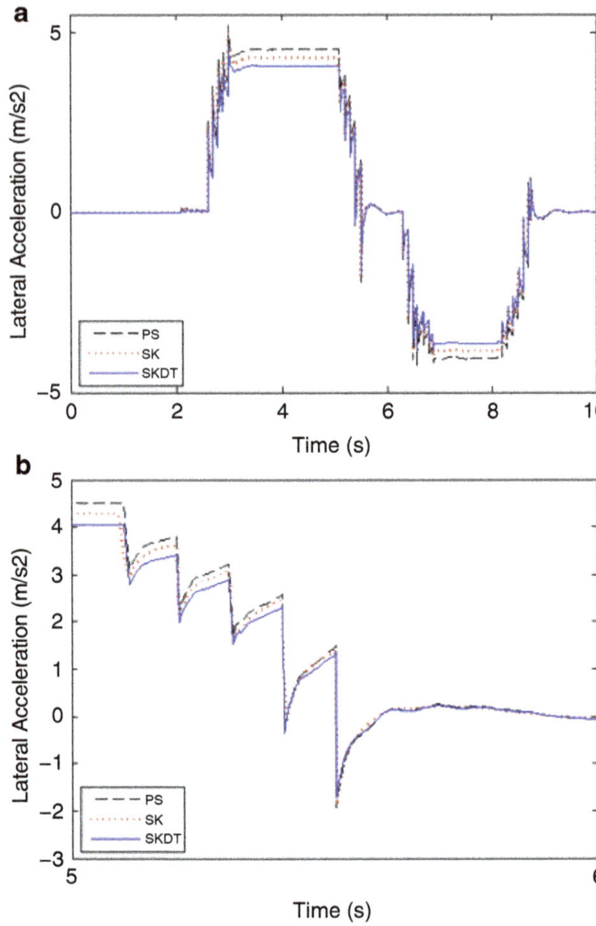

(faiD) for the steering input signal (Delta) and the road bank angle (β) according to the designed direct tilt control system was described in Sect. 7.3.2. In this subsection, the time domain analysis has also been presented to the vehicle body vertical, pitch angular, roll angular, lateral accelerations and vehicle rear right sprung mass vertical displacement.

7.3.1 Experiments of Ride Comfort in the Frequency Domain

Figures 7.20 and 7.21 show the frequency responses of the vehicle vertical, pitch angular and roll angular acceleration to road disturbance class C. For better visualization, the responses of vehicle vertical, pitch angular and roll angular

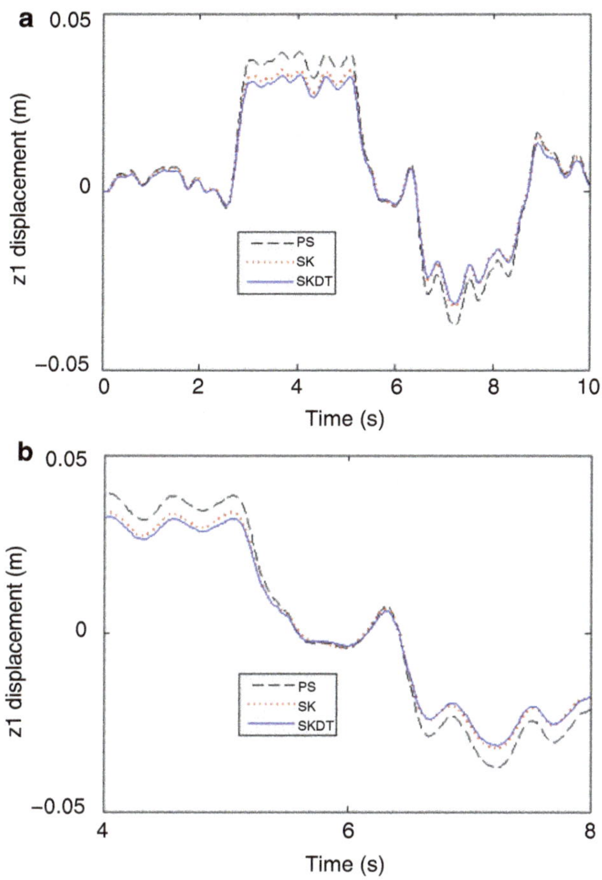

Fig. 7.12 The vehicle sprung mass m_1's vertical displacement for driving scenario four and road class C: (**a**) full trajectory and (**b**) short time span

accelerations at 0–10 Hz have been plotted separately in a short time span in Figs. 7.20a, 7.21a, and 7.22a, respectively.

The study shows that human beings are very sensitive to the vertical motion in the frequency range of 4–8 Hz [134]. So while designing a vehicle suspension system, sprung mass acceleration tolerance should be kept to a minimum in the frequency range mentioned above. From Figs. 7.20 and 7.21, it is quite evident that the vehicle body vertical acceleration and pitch angular acceleration of the SK and SKDT systems remained closer to 0 db than the PS system at a wide range of frequencies, especially at frequencies where humans are more sensitive to vibrations (Figs. 7.20a and 7.21a).

However the frequency responses of roll angular acceleration of the SK and SKDT systems have deteriorated compared to the PS system as shown in Fig. 7.22. Fortunately, the vehicle's roll angular acceleration only accounts for less than 20% of vehicle ride comfort, in terms of ride comfort index specified by ISO 2631-1 standard. Hence the vehicle ride comfort would not be affected significantly by the large roll angular acceleration. Moreover, the passengers would feel less lateral

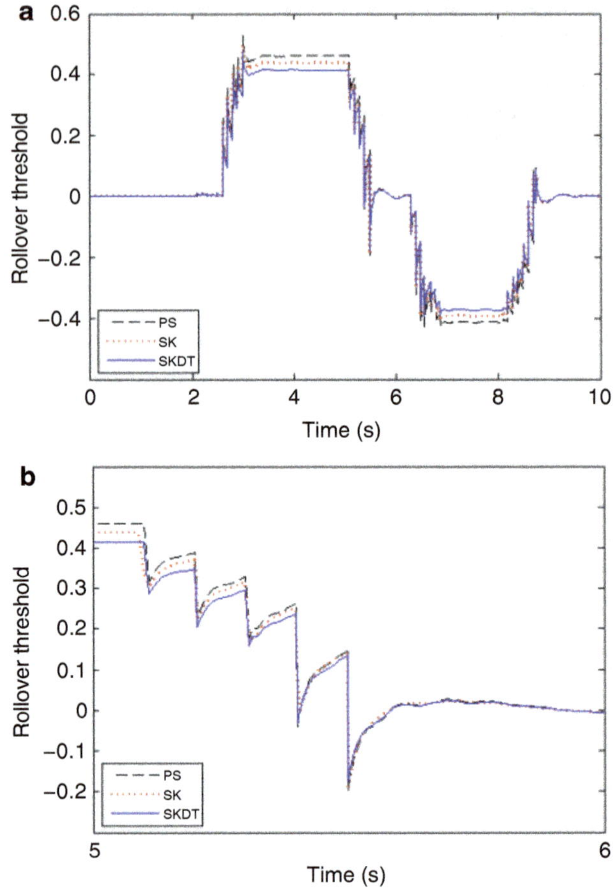

Fig. 7.13 The rollover threshold in driving scenario four and road class C: (a) full trajectory and (b) short time span

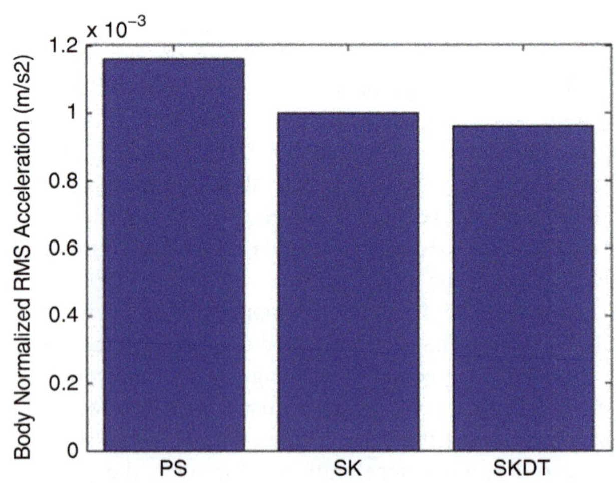

Fig. 7.14 Vehicle body vertical acceleration comparison

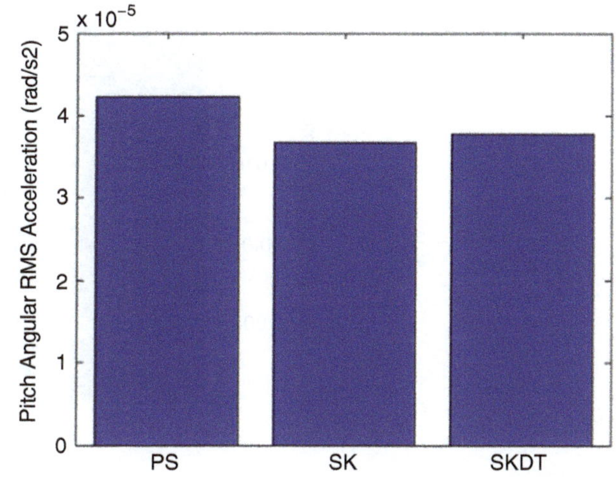

Fig. 7.15 Vehicle body pitch angular acceleration comparison

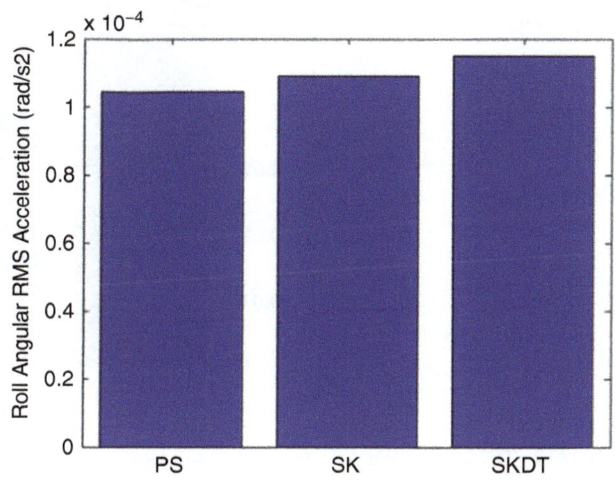

Fig. 7.16 Vehicle body roll angular acceleration comparison

acceleration; hence the ride comfort would be improved which is shown later in the time domain analysis (Fig. 7.23).

7.3.2 Experiments of Ride Comfort in the Time Domain

In this section, vehicle reactions to a driving scenario using the PS and the SKDT semi-active suspension systems are described. Figure 6.53a shows the response of the vehicle's desired tilting angle (faiD) for the steering input signal (Delta) and the road bank angle (β) according to the designed direct tilt control. The required

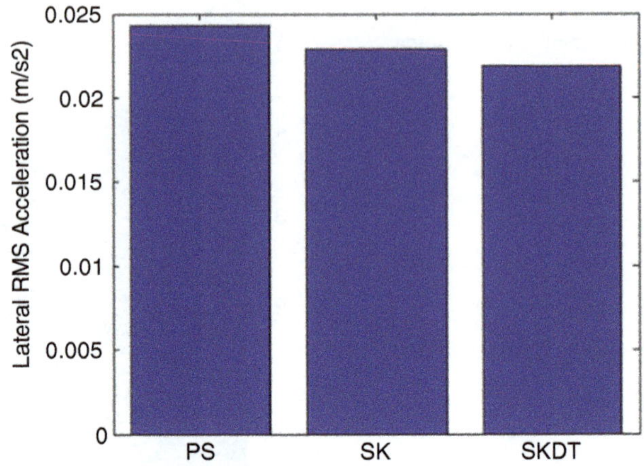

Fig. 7.17 Vehicle body lateral acceleration comparison

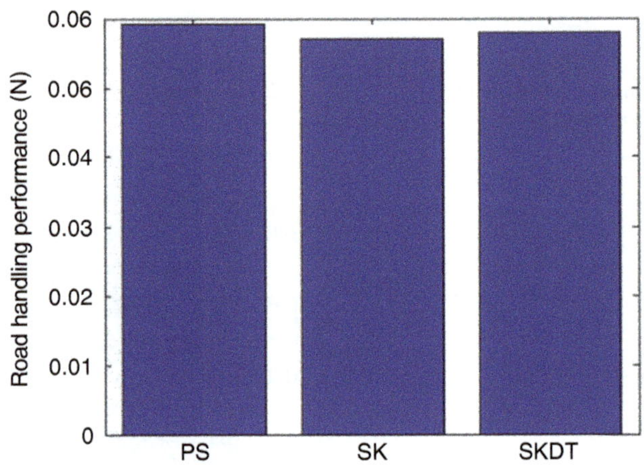

Fig. 7.18 Vehicle road handling performance comparison

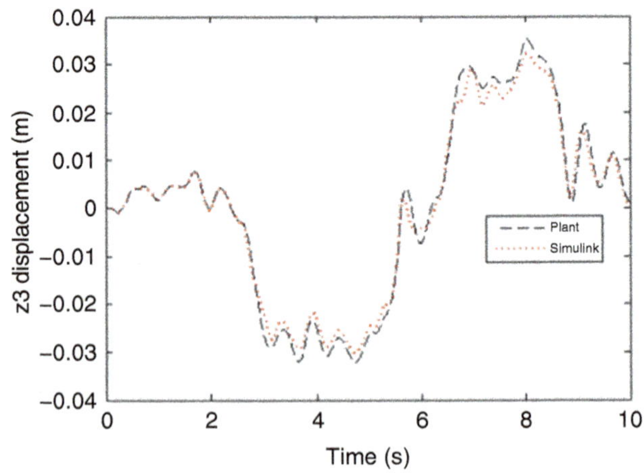

Fig. 7.19 The vehicle rear right sprung mass vertical displacement

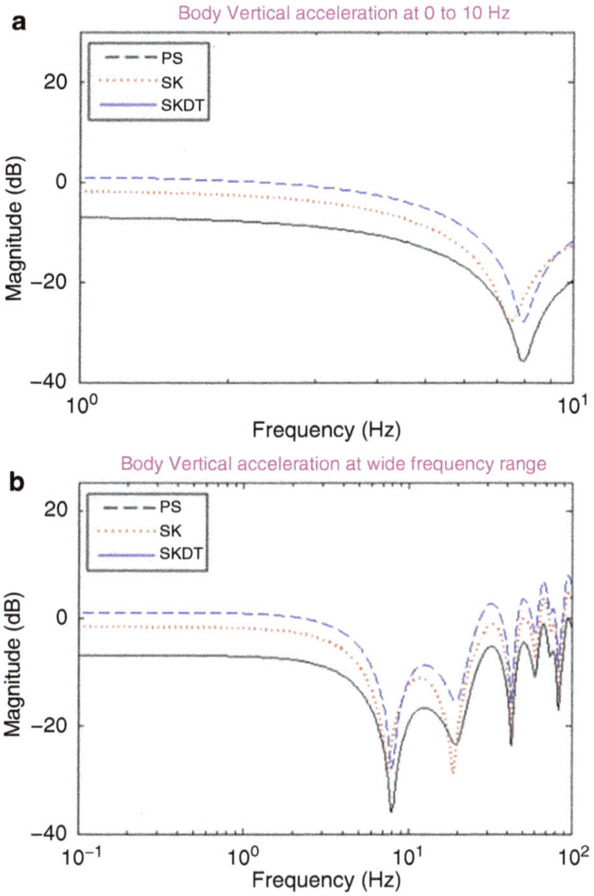

Fig. 7.20 The frequency response of vehicle body vertical acceleration: (**a**) at narrow frequency range and (**b**) at broad frequency range

actuator force to tilt the vehicle according to the desired tilting angle is depicted in Fig. 6.53b.

The time domain responses of the vehicle body vertical, pitch angular, roll angular, lateral accelerations and the vertical displacement of the vehicle rear right sprung mass m_3 for this driving scenario are shown in Figs. 7.24, 7.25, 7.26, 7.27, and 7.28, respectively. It is quite apparent that the displacement of sprung mass m_3, the vehicle vertical, pitch angular accelerations and the lateral accelerations of the SK and SKDT systems have decreased compared to the PS system. For better visualization, the responses of the displacement of sprung mass m_3, vehicle vertical, pitch angular roll angular and lateral acceleration and the maximum lateral acceleration value over gravity (a_y/g) have been plotted separately in a short time span.

If the body of the vehicle is parallel with the ground surface, its rollover threshold becomes a function of the vehicle's track width (t) and the centre of gravity height (h). The maximum allowable rollover threshold of this vehicle is

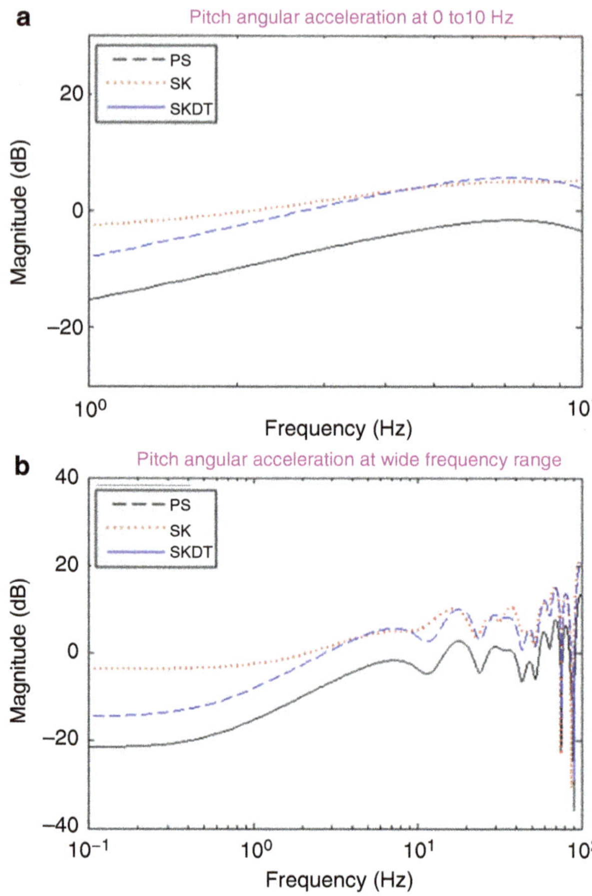

Fig. 7.21 The frequency response of vehicle body pitch angular acceleration: (**a**) at narrow frequency range and (**b**) at broad frequency range

1.4578. The rollover threshold defines the maximum lateral acceleration value over gravity (a_y/g) that the vehicle would be able to reach before tipping over. From Fig. 7.29a, b, it is clear that (a_y/g) never exceeds the maximum allowable rollover threshold for this vehicle. Hence the vehicle remains stable throughout the trajectory of this driving scenario.

Compared to the passive suspension system, the normalized RMS body vertical acceleration (NV) of the SK and SKDT suspension systems have decreased by 14.21% and 17.67%, respectively, as shown in Fig. 7.30. The normalized RMS pitch angular accelerations (NP) of the SK and SKDT suspension systems decreased by 12.27% and 10.88%, respectively (Fig. 7.31).

However Fig. 7.32 shows that the normalized RMS roll angular acceleration (NR) of the SK and SKDT systems have increased by 0.34% and 10%. Compared to the PS system, the NL (normalized lateral acceleration) of the SK and SKDT suspension systems have decreased by 5.26% and 9%, respectively (Fig. 7.33). As shown in Fig. 7.34, the road handling performance of the SK and SKDT systems

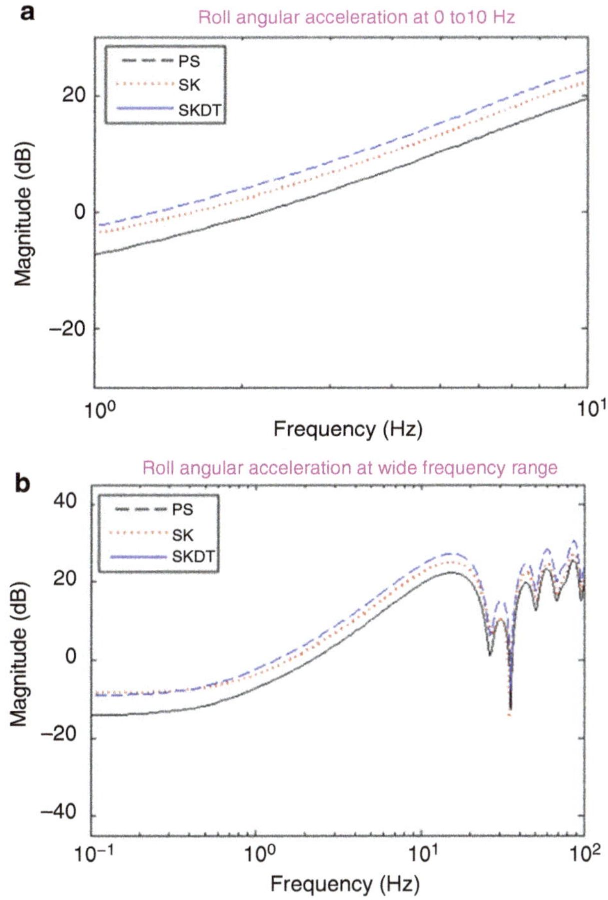

Fig. 7.22 The frequency response of vehicle body roll angular acceleration: (**a**) at narrow frequency range and (**b**) at broad frequency range

has decreased by only 2.66% and 2% compared to the passive suspension system which is within the acceptable range. Hence, the SK and SKDT suspension controllers can effectively improve vehicle ride comfort while the vehicle is moving on road class C.

7.4 Conclusions

This chapter described the conclusions reached from the experimental analysis of the full car model along with the Quanser quarter-car suspension plant that was carried out to investigate the proposed skyhook control suspension system and the direct tilt control system. The vehicle performance analysis using the Quanser quarter-car suspension plant at the front left and rear right wheel of the full car model was presented separately in two different sections. It has been observed that

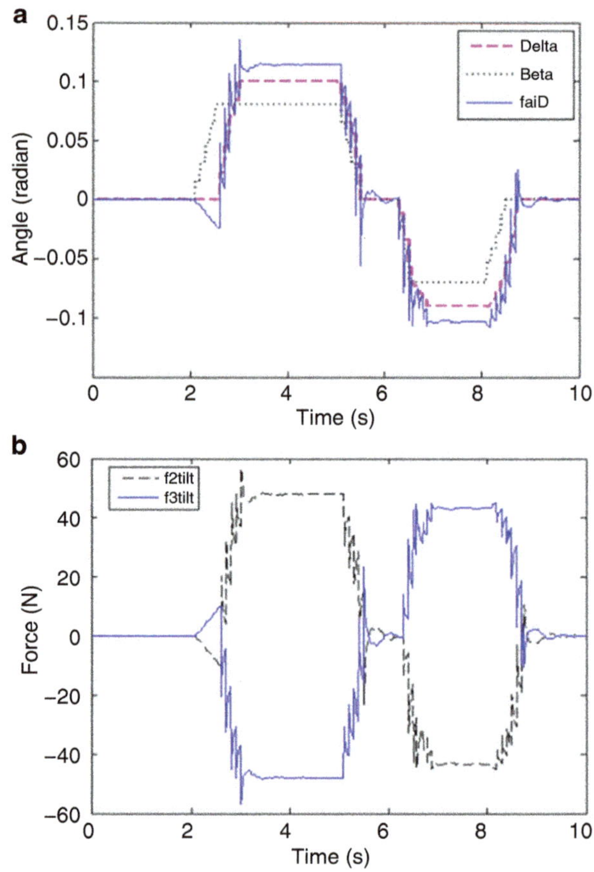

Fig. 7.23 The response of steering and bank angle in driving scenario four and road class C: (**a**) desired tilting angle and (**b**) required actuator force

the results of the two environments are slightly different. The reason is the distances from CG to the front and rear wheels are different for the full car model.

The SK and SKDT control systems were applied to the dynamic model of the full car. It takes the road bank angle into consideration to evaluate the performances of the controllers. The behaviour of the vehicle body vertical acceleration, pitch angular acceleration and roll angular acceleration were evaluated in the frequency domain. While travelling on all the road profiles, it has been ascertained that the vehicle body vertical acceleration and pitch angular acceleration of both the SK and SKDT systems remained closer to 0 db than the PS system at a wide range of frequencies, especially at frequencies where humans are more sensitive to vibrations.

Time domain analysis was also carried out to investigate the performance of the systems. NV (normalize body vertical acceleration) and NP (normalized pitch angular acceleration) have been calculated, and it has been found that both the SK and SKDT systems significantly reduced the vehicle body vertical acceleration and pitch angular acceleration in all instances compared to the PS system. However

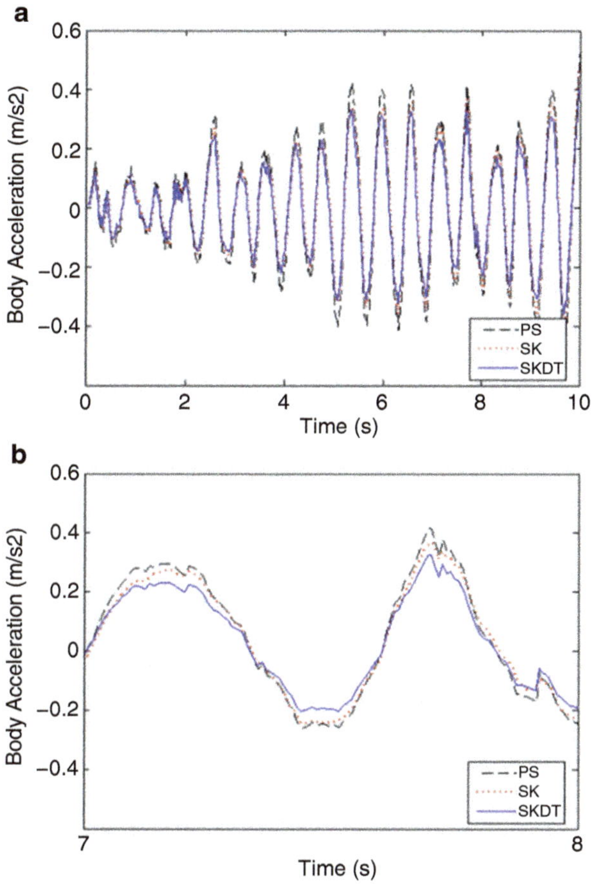

Fig. 7.24 The vehicle body vertical acceleration for driving scenario four and road class C: (**a**) full trajectory and (**b**) short time span

the normalized RMS of the roll angular acceleration (NR) of the SKDT system increased significantly with the tilting action. Fortunately, the vehicle's roll angular acceleration only accounts for less than 20% of vehicle ride comfort, in terms of ride comfort index specified by ISO 2631-1 standard. Hence the vehicle ride comfort would not be affected significantly by the large roll angular acceleration. Moreover, the appropriate tilting action reduced the lateral acceleration experienced by passengers and improved the vehicle ride comfort, as the NL (normalized lateral acceleration) of the SK and SKDT suspension systems have decreased compared to the PS system. The road handling performance of both suspension systems has also been evaluated, and it has been found that the performance decreased slightly compared to the passive suspension system which is within the acceptable range. The stability of the resulting closed-loop system was then investigated.

The response of the vehicle's desired tilting angle (faiD) for the corresponding steering input signal (Delta) and the road bank angle (β) has been shown. The required actuator force to tilt the vehicle according to the desired tilting angle has also been

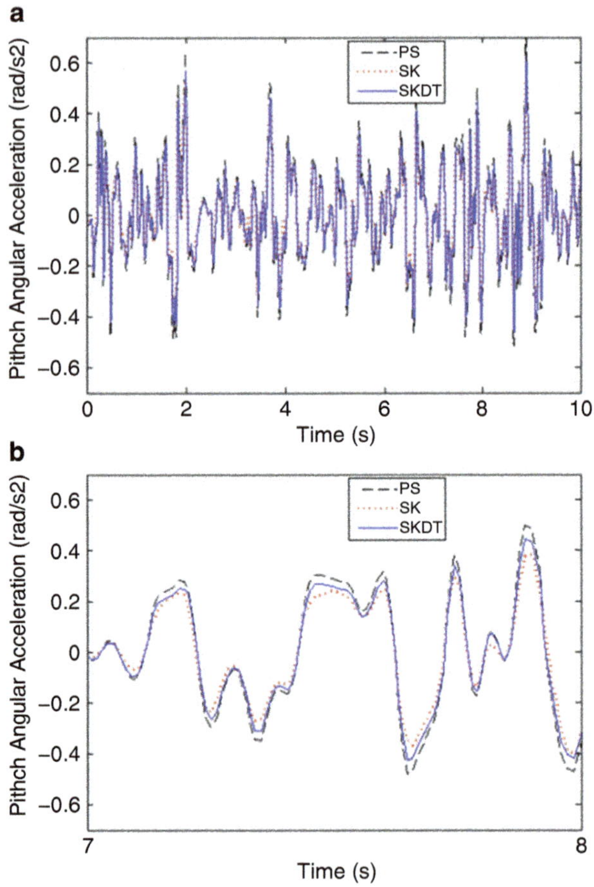

Fig. 7.25 The pitch angular acceleration for driving scenario four and road class C: (**a**) full trajectory and (**b**) short time span

observed. Overall it has been realized that the proposed skyhook control system on its own can improve the vehicle ride comfort keeping the road handling performance within an acceptable range. The SKDT system shows that this control strategy is capable of tilting a vehicle inward which would act against the lateral acceleration result of the steering manoeuvres. The designed SKDT system was capable of maintaining the system's tilt state by keeping the vehicle stable and safe from rollover.

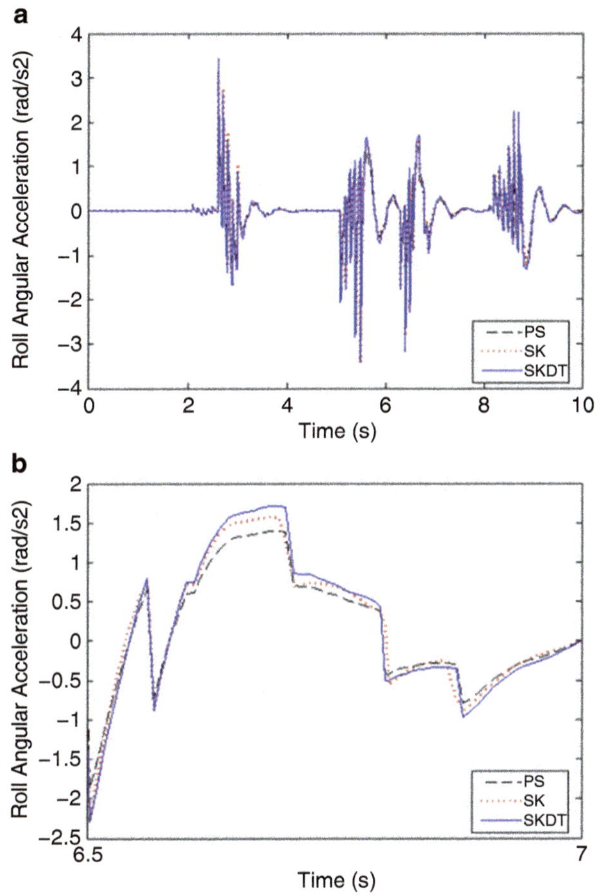

Fig. 7.26 The roll angular acceleration for driving scenario four and road class C: (**a**) full trajectory and (**b**) short time span

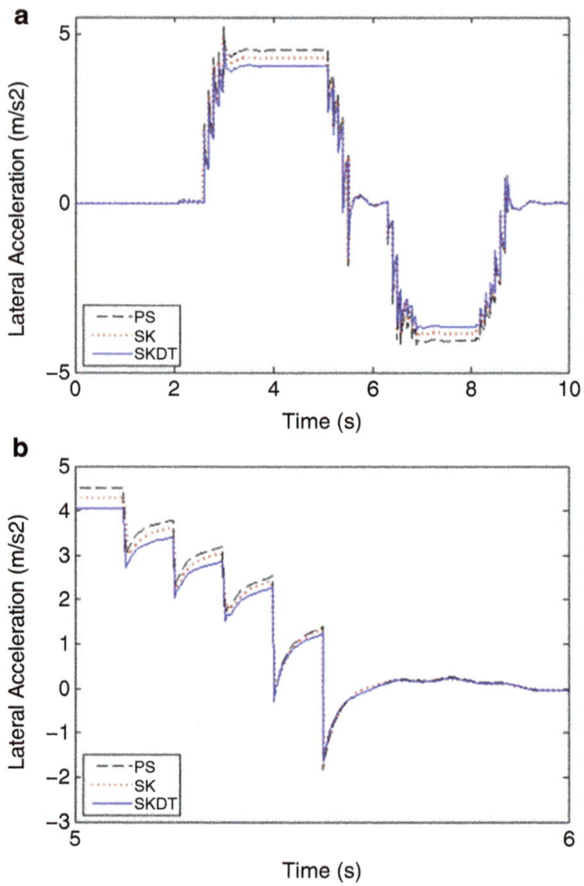

Fig. 7.27 The lateral acceleration for driving scenario four and road class C: (**a**) full trajectory and (**b**) short time span

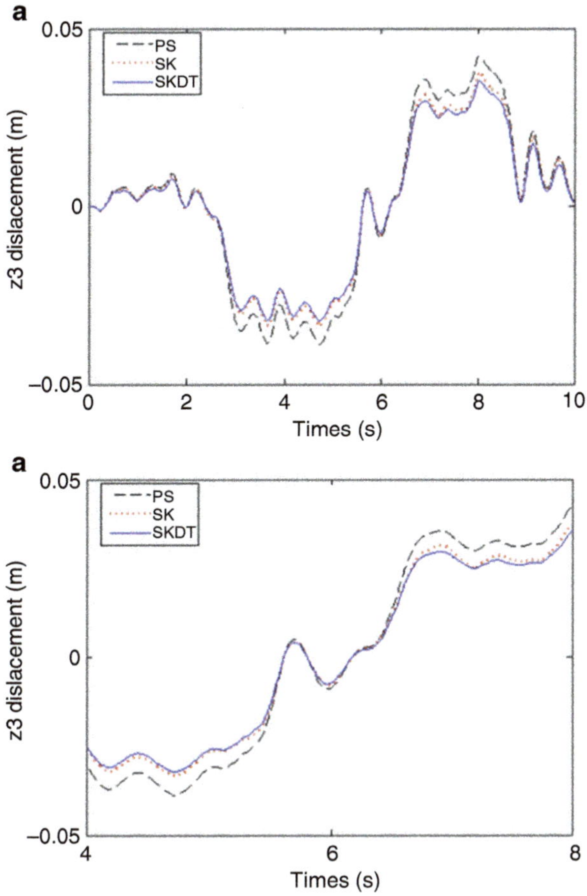

Fig. 7.28 The vehicle sprung mass m_3's vertical displacement for driving scenario four and road class C: (**a**) full trajectory and (**b**) short time span

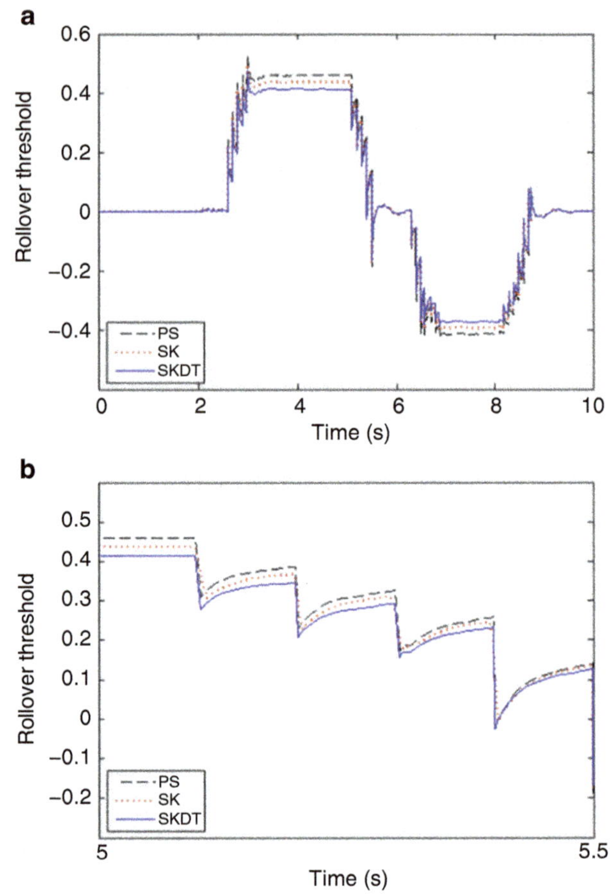

Fig. 7.29 The rollover threshold in driving scenario four and road class C: (**a**) full trajectory and (**b**) short time span

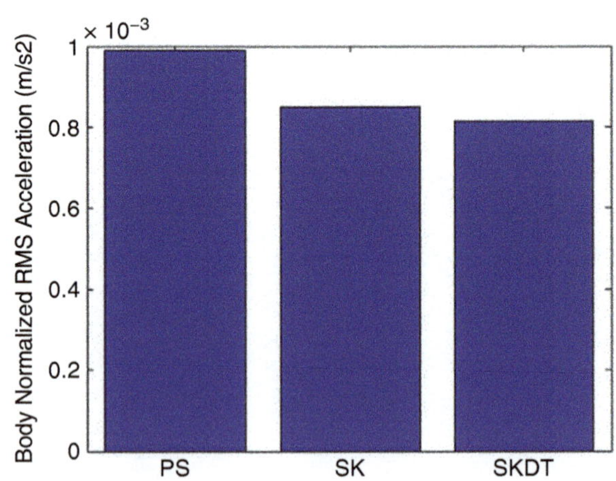

Fig. 7.30 Vehicle body vertical acceleration comparison

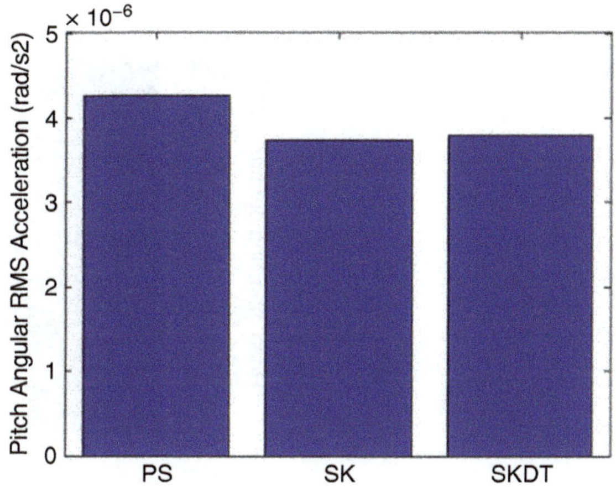

Fig. 7.31 Vehicle body pitch angular acceleration comparison

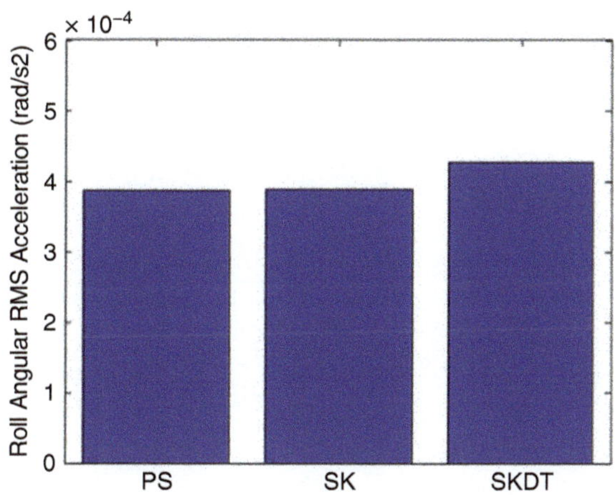

Fig. 7.32 Vehicle body roll angular acceleration comparison

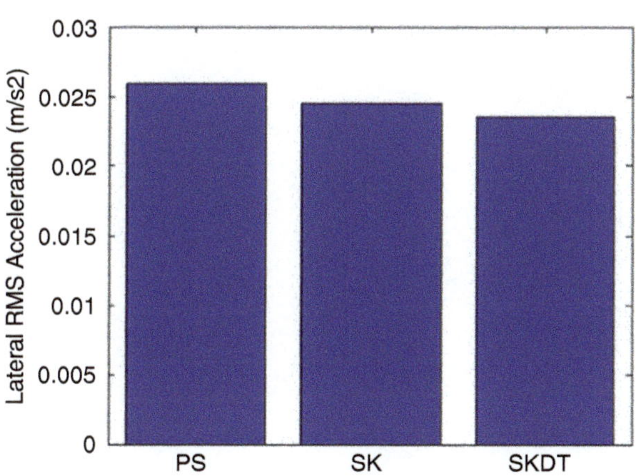

Fig. 7.33 Vehicle body lateral acceleration comparison

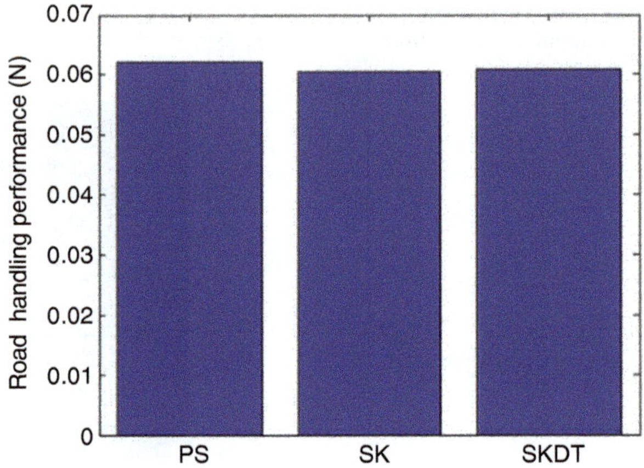

Fig. 7.34 Vehicle road handling performance comparison

Chapter 8
Conclusions and Recommendations

Abstract This manuscript began by considering two key elements as the main objectives of this book. The first goal was to develop a semi-active control strategy that would reduce the unwanted effects of various road conditions and driving manoeuvres on the vehicle body and the passengers. The second objective was to design and study the performance of a semi-active suspension system suitable for the standard passenger vehicle and to improve the cornering performance of the vehicle. In this chapter, the achievement of the research has been discussed with the recommendation for the future work.

8.1 Introduction

In this manuscript, various types of semi-active control approaches for vehicle suspension systems have been studied. A new modified skyhook control strategy with adaptive skyhook gain has been presented which is claimed as a novel idea by the reviewers in the International Journal of Vehicle Mechanics and Mobility titled, "Vehicle System Dynamics".

An extensive literature review has been done on both academic research and industrial advancement of vehicle tilting technology to improve cornering performance. But to date, none of them considered the road bank angle in the control system design and the modelling of the dynamic model of the standard passenger tilting vehicles. This study designed and developed a new analytical full-vehicle model incorporating the road bank angle in the vehicle dynamics. A direct tilt control method along with the proposed modified skyhook-controlled closed-loop feedback system was developed to control the tilt action of the vehicle. This system improved the performance of the vehicle during cornering with little or no skidding. This was achieved by using a new approach of tilting the standard passenger vehicle inward during cornering considering the road bank angle along with the steering angle, lateral position acceleration, yaw rate and velocity of the vehicle.

S. Kashem et al., *Vehicle Suspension Systems and Electromagnetic Dampers*,
Springer Tracts in Mechanical Engineering, DOI 10.1007/978-981-10-5478-5_8

8.2 Overview of the Study

After a brief introduction on automotive suspension systems in the first chapter, an extensive literature review was done on many robust and optimal control approaches or algorithms. This included linear time-invariant H-infinity control (LTIH), linear parameter-varying control (LPV) and model-predictive controls (MPC). Five widely known control approaches were reviewed more deeply, namely, the linear quadratic regulator and linear quadratic Gaussian, sliding mode control, fuzzy and neuro-fuzzy control and skyhook and groundhook approaches. It was found that the skyhook control strategy is the most widely used among all other control strategies due to its simplicity for practical implementation. But still there is a great scope of work to be done to modify the skyhook control strategy to achieve better performance. Different types of damper technologies have been also discussed in this manuscript, and it has been ascertained that the linear electromagnetic damper is best for the semi-active suspension system due to its fast response time which is better than the best hydraulic device. A brief literature review on automotive tilting technology was done in this manuscript which suggests that a direct tilting method needs to be developed to tilt the standard passenger vehicle inward during cornering considering the road bank angle.

The vehicle suspension system has been categorized and discussed briefly. It has been explained that the semi-active suspension system is the most suitable for investigation in this manuscript. A brief description of the quarter-car model has been given as well as an explanation of the motion equations used in the model. The high- and low-bandwidth suspension systems have also been discussed. As there is no requirement for a static load force in this suspension model, a two degree of freedom HBC semi-active suspension system was used to investigate different semi-active control algorithms in this manuscript. An extensive comparison of quarter-car models extracted from the study of various researchers has been presented.

A brief discussion on a proposed modified skyhook control approach, the optimal skyhook control of Nguyen et al. [51], modified skyhook control of Bessinger et al. [15] and continuous skyhook control of Karnopp et al. [14] was presented. A road profile has been generated to study the performance of the different controllers. The two degrees of freedom quarter-car model has been simulated to compare the controllers' performances. The Quanser quarter-car suspension plant has been also used to compare the performance of the controllers in an experimental environment. From the above simulation and experimental analysis, it can be concluded that the proposed modified skyhook-controlled closed-loop feedback system provides the best performance of those investigated because it decreases sprung-mass acceleration to a large extent compared to the passive system and other skyhook controllers described in the literature. It also keeps road-handling performance within the acceptable range. These models have also been evaluated in terms of human vibration perception and admissible acceleration levels based on ISO 2631.

This manuscript also presents a methodology on how to integrate the proposed skyhook control in a full-car model to improve ride comfort and handling via a semi-active suspension system. It demonstrates the direct tilt control strategy which can tilt a vehicle inward to act against the roll movement due to steering manoeuvres as well as road bank angle. A technique to determine the vehicle rollover propensity has also been described. The road profile and four driving scenarios have also been briefly discussed in this chapter which forms a basis for the analysis described in the next following chapters. To compare and evaluate the results of the simulation environment and experimental setup, three criteria have been set. The ride comfort of the vehicle will be evaluated by comparing the results of the normalized RMS value of vehicle body vertical acceleration, vehicle pitch angular acceleration, roll angular acceleration and lateral acceleration. Evaluation of road-handling performance will be done by comparing the tyre forces normalized to the static forces that acts on the vehicle's wheels.

The SK and SKDT systems were applied to the dynamic model of a full car which considers the road bank angle. At first, the responses of the dynamic model of the full car on different road classes A, B and C defined by ISO8608 and also a combined road profile have been observed. In this section, PS and SK systems were analysed separately. The behaviour of the vehicle body vertical acceleration and pitch angular acceleration was examined in the frequency domain. While travelling on all the road profiles, it has been ascertained that the vehicle body vertical acceleration and pitch angular acceleration of the SK system remain closer to 0 db than the PS system at a wide range of frequencies, particularly in the region where humans are more sensitive to vibrations.

The time domain analysis was also carried out on the full-car model using the PS and the SK system separately. In this subsection, vertical displacement of the vehicle front left sprung mass m_1, vehicle body vertical acceleration and pitch angular acceleration have been assessed. NV (normalized body vertical acceleration) and NP (normalized pitch angular acceleration) have also been calculated, and it has been found that the SK system significantly reduced the vehicle body vertical acceleration and pitch angular acceleration in all instances compared to the PS system. However the road-handling performance of the SK-controlled suspension system decreased slightly compared to the passive suspension system. But the SK system's weighted RMS acceleration value of the sprung mass obtained in this simulation was always within the acceptable range of international code ISO 2631 [143]. Since the vehicle was traversing on a straight road, the differences in roll angular and lateral acceleration of both systems were negligible in this analysis.

Subsequently, the simulations of the dynamic model of a full car with different driving scenarios [described in Sect. (5.3)] were carried out using the SKDT system. Vehicle reaction to the driving scenarios of both the PS and the SKDT semi-active suspension systems has been described in this section. The response of the vehicle's desired tilting angle (faiD) for the corresponding steering input signal (Delta) and the road bank angle (β) has been shown. The required actuator force to tilt the vehicle according to the desired tilting angle was also observed.

Here the time domain responses of the vertical displacement of the vehicle front left sprung mass $m_{1,}$ vehicle body vertical, pitch angular, roll angular and lateral acceleration for each driving scenario have been shown, and it was quite evident that the displacement of sprung mass m_1, the vehicle vertical, pitch angular accelerations and the lateral acceleration of the SKDT system has decreased compared to the PS system. Compared to the passive suspension system, the NV and NP of the SKDT suspension system have decreased significantly. However the normalized RMS of the roll angular acceleration (NR) of the SKDT system was increased significantly for the tilting action. Fortunately, the roll angular acceleration of vehicle only accounts for $<20\%$ of vehicle ride comfort, in terms of ride comfort index specified by the ISO 2631-1 standard. Hence vehicle ride comfort would not be affected significantly by the large roll angular acceleration. Moreover, the appropriate tilting action reduces the lateral acceleration experienced by passengers and improved the vehicle ride comfort, as the NL (normalized lateral acceleration) of the SKDT suspension system was decreased by 10.01% compared to the PS system. The stability of the resulting closed-loop system was then investigated.

Road class C and driving scenario four were used to evaluate the performance of the PS, SK and SKDT systems. It has been realized that the proposed skyhook-controlled closed-loop feedback system on its own can improve vehicle ride comfort keeping the road-handling performance within an acceptable range. The SKDT system shows that this control strategy is capable of tilting a vehicle inward which would act against the lateral acceleration result of the steering manoeuvres. The designed SKDT system was capable of maintaining the system's tilt state by keeping the vehicle stable and safe from rollover. This manuscript has indicated the potential of the SKDT suspension system in improving cornering performances of the vehicle and paves the way of a future work on a vehicle's integrated system for chassis control.

The vehicle performance analysis using the Quanser quarter-car suspension plant at front left and rear right wheel of the full-car model is presented separately in two different sections. It has been observed that the results of the two environments are slightly different. The reason is the distances from C.G. to the front and rear wheel is different for the full-car model.

The SK and SKDT control systems were applied to the dynamic model of the full-car model which considers the road bank angle to evaluate the performances of the controllers. The behaviour of the vehicle body vertical acceleration, pitch angular acceleration and roll angular acceleration was examined in the frequency domain. While travelling on all the road profiles, it has been ascertained that the vehicle body vertical acceleration and pitch angular acceleration of both the SK and SKDT systems remained closer to 0 db than the PS system at a wide range of frequencies, particularly in the region where humans are more sensitive to vibrations.

The time domain analysis was also carried out to investigate the performance of the systems. NV (normalized body vertical acceleration) and NP (normalized pitch angular acceleration) have also been calculated, and it has been found that both the SK and SKDT system significantly reduced the vehicle body vertical acceleration and pitch angular acceleration in all instances compared to the PS system. However the normalized RMS of the roll angular acceleration (NR) of the SKDT system increased significantly for the tilting action. Fortunately, the roll angular acceleration of vehicle only accounts for <20% of vehicle ride comfort, in terms of ride comfort index specified by the ISO 2631-1 standard. Hence the vehicle ride comfort would not be affected significantly by the large roll angular acceleration. Moreover, the appropriate tilting action reduces the lateral acceleration experienced by passengers and improved the vehicle ride comfort, as the NL (normalized lateral acceleration) of the SK and SKDT suspension system decreased compared to the PS system. The road-handling performances of both suspension systems have also been evaluated, and it has been found that the performance decreased slightly compared to the passive suspension system which is within the acceptable range. The stability of the resulting closed-loop system was then investigated.

The response of the vehicle's desired tilting angle (faiD) for the corresponding steering input signal (Delta) and the road bank angle (β) has been shown. The required actuator force to tilt the vehicle according to the desired tilting angle had also been observed. Overall it has been realized that the proposed skyhook control system on its own can improve the vehicle ride comfort keeping the road-handling performance within an acceptable range. The SKDT system shows that this control strategy is capable of tilting a vehicle inward which would act against the lateral acceleration resulting from the steering manoeuvres. The designed SKDT system was capable of maintaining the system's tilt state by keeping the vehicle stable and safe from rollover.

This manuscript presents a novel idea on how to integrate vehicle ride comfort and handling control via semi-active suspensions. It demonstrated that the integrated suspension control strategy can tilt a vehicle inward to act against the roll moment due to steering manoeuvres.

8.3 Recommendations for Future Study

This manuscript has indicated the potential of the SKDT suspension system in improving cornering performances of the vehicle and paves the way for future work on vehicle's integrated system for chassis control. Based on my completed research work on semi-active dampers and semi-active controlled suspension systems, the following research is recommended for further study.

Development of a microcontroller for the proposed modified skyhook-controlled closed-loop feedback system and the direct tilt control method is an important aspect of future work. Theoretical and experimental studies of the control methods have already been done.

Development of a commercial semi-active suspension system with linear electromagnetic damper would be a potential work.

Real-time testing of the developed damper control strategies on an actual standard passenger vehicle suspension system and testing on standard test tracks is recommended. Comparing the real-time test results with the results obtained through numerical simulation using CarSim or Adams/Car would be beneficial as well.

Appendix A

Simulink Model

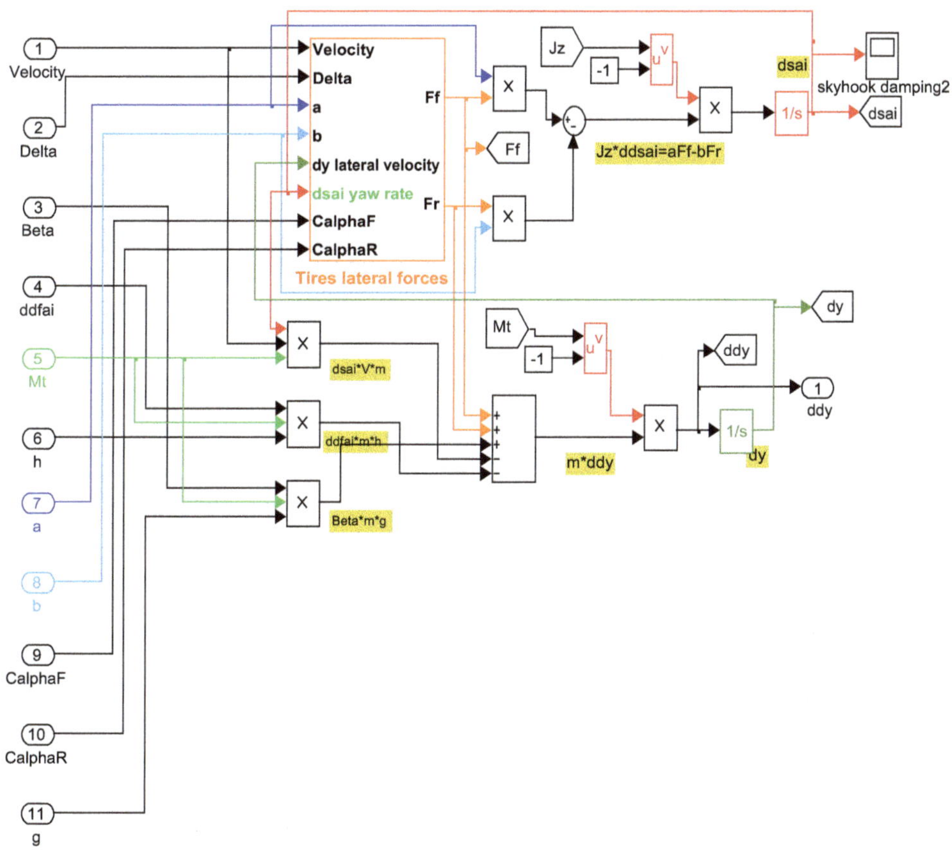

Fig. A1 Determine lateral position acceleration

S. Kashem et al., *Vehicle Suspension Systems and Electromagnetic Dampers*,
Springer Tracts in Mechanical Engineering, DOI 10.1007/978-981-10-5478-5

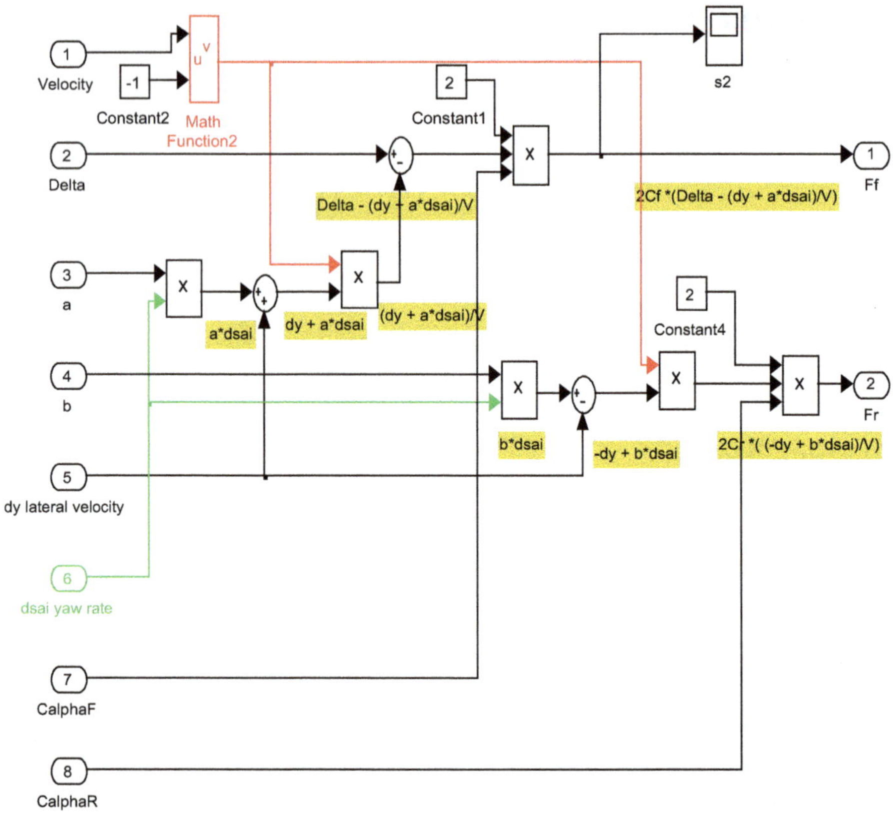

Fig. A2 Determine the front and rear tires lateral forces

Appendix B

MATLAB Code for Frequency Domain Analysis of Full Car Model

```
clc
clear all
close all

load('C:\Users\s\Dropbox\Full      car      cornering      160512
\roadprofileClassABC20sEach280512.mat')

SpecifiedStopTime = 60;

 RoadSteerBank280512;
 gcs;
 set_param('RoadSteerBank280512', 'StopTime', 'SpecifiedStopTime');
 sim('C:\Users\s\Dropbox\Full      car      cornering      160512
\RoadSteerBank280512.mdl')

% set_param('RoadSteerBank280512', 'SimulationCommand', 'start');
% pause(SpecifiedStopTime);

% roadprofile = roadprofileClassA;
% roadprofile = roadprofileClassB;
roadprofile = roadprofileClassC;
% roadprofile = roadprofileClassABC20sEach280512; %change simula-
tion time of the full model = dsaiDes;
```


S. Kashem et al., *Vehicle Suspension Systems and Electromagnetic Dampers*,
Springer Tracts in Mechanical Engineering, DOI 10.1007/978-981-10-5478-5

```
% Delta = SteerZero;
% Beta = BetaZero;

% Delta = Steer1;
% Beta = BetaZero;
%
% Delta = SteerZero;
% Beta = Beta1;
%
% Delta = Steer1;
% Beta = Beta1;
%
Delta = Steer2;
Beta = Beta2;

%% ****************************************
%% First of all, chane the stop time to 10 then
%% **********************************************
full_carstabilizedcModelz1z2z3z4usedSfunc160512;
gcs;
set_param('full_carstabilizedcModelz1z2z3z4usedSfunc160512',
'StopTime', 'SpecifiedStopTime');
%    set_param('full_carstabilizedcModelz1z2z3z4usedSfunc160512',
'SimulationCommand', 'start');

full_carstabilizedcModelz1z2z3z4usedSfunc160512skyhook;
gcs;
set_param('full_carstabilizedcModelz1z2z3z4usedSfunc160512skyhook',
'StopTime', 'SpecifiedStopTime');
% set_param('full_carstabilizedcModelz1z2z3z4usedSfunc160512skyhook',
'SimulationCommand', 'start');

full_carstabilizedcModelz1z2z3z4usedSfunc160512DTC;
gcs;
set_param('full_carstabilizedcModelz1z2z3z4usedSfunc160512DTC',
'StopTime', 'SpecifiedStopTime');
% set_param('full_carstabilizedcModelz1z2z3z4usedSfunc160512DTC',
'SimulationCommand', 'start');
% pause(SpecifiedStopTime);

sim('C:\Users\s\Dropbox\Full car cornering 160512
\full_carstabilizedcModelz1z2z3z4usedSfunc160512.mdl')
sim('C:\Users\s\Dropbox\Full car cornering 160512
\full_carstabilizedcModelz1z2z3z4usedSfunc160512skyhook.mdl')
```

```
%     sim('C:\Users\s\Dropbox\Full     car     cornering     160512
\full_carstabilizedcModelz1z2z3z4usedSfunc160512DTC.mdl')

% Now resample the outputs and compare
SkyhookTimeSeriesOutput  =  timeseries(FullVehicleSkyhook.signals.
values(:,2),FullVehicleSkyhook.time);
ReSampledSkyhook2 = resample(SkyhookTimeSeriesOutput,FullVehicle.
time);

SkyhookTimeSeriesOutput  =  timeseries(FullVehicleSkyhook.signals.
values(:,4),FullVehicleSkyhook.time);
ReSampledSkyhook4 = resample(SkyhookTimeSeriesOutput,FullVehicle.
time);

SkyhookTimeSeriesOutput  =  timeseries(FullVehicleSkyhook.signals.
values(:,6),FullVehicleSkyhook.time);
ReSampledSkyhook6 = resample(SkyhookTimeSeriesOutput,FullVehicle.
time);

SkyhookTimeSeriesOutput  =  timeseries(FullVehicleSkyhook.signals.
values(:,12),FullVehicleSkyhook.time);
ReSampledSkyhook12       =       resample(SkyhookTimeSeriesOutput,
FullVehicle.time);

SkyhookTimeSeriesOutput  =  timeseries(FullVehicleSkyhook.signals.
values(:,13),FullVehicleSkyhook.time);
ReSampledSkyhook13       =       resample(SkyhookTimeSeriesOutput,
FullVehicle.time);

SkyhookTimeSeriesOutput  =  timeseries(FullVehicleSkyhook.signals.
values(:,14),FullVehicleSkyhook.time);
ReSampledSkyhook14       =       resample(SkyhookTimeSeriesOutput,
FullVehicle.time);

SkyhookTimeSeriesOutput  =  timeseries(FullVehicleSkyhook.signals.
values(:,15),FullVehicleSkyhook.time);
ReSampledSkyhook15       =       resample(SkyhookTimeSeriesOutput,
FullVehicle.time);

SkyhookTimeSeriesOutput  =  timeseries(FullVehicleSkyhook.signals.
values(:,16),FullVehicleSkyhook.time);
ReSampledSkyhook16       =       resample(SkyhookTimeSeriesOutput,
FullVehicle.time);
```

```
SkyhookTimeSeriesOutput = timeseries(FullVehicleSkyhook.signals.
values(:,18),FullVehicleSkyhook.time);
ReSampledSkyhook18 = resample(SkyhookTimeSeriesOutput,
FullVehicle.time);

SkyhookTimeSeriesOutput = timeseries(FullVehicleSkyhook.signals.
values(:,20),FullVehicleSkyhook.time);
ReSampledSkyhook20 = resample(SkyhookTimeSeriesOutput,
FullVehicle.time);

SkyhookTimeSeriesOutput = timeseries(FullVehicleSkyhook.signals.
values(:,21),FullVehicleSkyhook.time);
ReSampledSkyhook21 = resample(SkyhookTimeSeriesOutput,
FullVehicle.time);

SkyhookTimeSeriesOutput = timeseries(FullVehicleSkyhook.signals.
values(:,26),FullVehicleSkyhook.time);
ReSampledSkyhook26 = resample(SkyhookTimeSeriesOutput,
FullVehicle.time);

 close all

% %% %% Finally determine the frequencies of models

NFFT = 2^nextpow2(length(FullVehicle.time)); % Next power of
2 from length of the data
NFFT = 2^(nextpow2(length(FullVehicle.time))+2);
f = 1/((FullVehicle.time(22))*2)*linspace(0,1,NFFT/2+1);
f1 = [-f(end-1:-1:2), f];

% %% %% % body acceleration
fTmp = fft((FullVehicle.signals.values(:,1)),NFFT)/length
(FullVehicle.time);
fVib2 = 20*log10(2*abs(fftshift(fTmp)));

fTmp3 = fft((ReSampledSkyhook2.Data),NFFT)/length
(ReSampledSkyhook2.time);
fVib3 = 20*log10(2*abs(fftshift(fTmp3)));

fTmp5 = fft((FullVehicle.signals.values(:,2)),NFFT)/length
(FullVehicle.time);
fVib5 = 20*log10(2*abs(fftshift(fTmp5)));
```

```
figure(51)
 semilogx(f1,fVib2(1:length(f1)),'k')
 hold on
 semilogx(f1,fVib3(1:length(f1)),'Color',[0.84                0.16
0],'LineStyle',':','LineWidth',2)
    hold on
 semilogx(f1,fVib5(1:length(f1)),'b','LineStyle','--
','LineWidth',1),

 xlabel(['Frequency (Hz)',sprintf('\n'),'[a]'], 'FontSize',13)
 ylabel('Magnitude (dB)','FontSize',13),axis([1e0 1e1 -25 25]),
 set(gca,'YTick',[-40 -20 0 20],'FontSize',13),
  legend(('PS','SK','SKDT'),'Location','NorthWest','FontSize',13)
   title('Body Vertical acceleration at 0 to 10 Hz','Color',[1 0
1],'FontSize',10)

  figure(52)
 semilogx(f1,fVib2(1:length(f1)),'k')
 hold on
 semilogx(f1,fVib3(1:length(f1)),'Color',[0.84                0.16
0],'LineStyle',':','LineWidth',2)
 hold on
 semilogx(f1,fVib5(1:length(f1)),'b','LineStyle','--
','LineWidth',1),

 xlabel(['Frequency  (Hz)',sprintf('\n'),'[b]'],  'FontSize',13),
ylabel('Magnitude (dB)','FontSize',13),axis([1e-1 1e2 -40 25]),
 set(gca,'YTick',[-40 -20 0 20],'FontSize',13),
  legend(('PS','SK','SKDT'),'Location','NorthWest','FontSize',13)
     title('Body  Vertical  acceleration  at  wide  frequency
range','Color',[1 0 1],'FontSize',10)

 % %% %% pitch

fTmp     =     fft((FullVehicle.signals.values(:,5)),NFFT)/length
(FullVehicle.time);
fVib2 = 20*log10(2*abs(fftshift(fTmp)));

fTmp3         =         fft((ReSampledSkyhook6.Data),NFFT)/length
(ReSampledSkyhook6.time);
fVib3 = 20*log10(2*abs(fftshift(fTmp3)));

fTmp5     =     fft((FullVehicle.signals.values(:,6)),NFFT)/length
(FullVehicle.time);
fVib5 = 20*log10(2*abs(fftshift(fTmp5)));
```

```
figure(53)
 semilogx(f1,fVib2(1:length(f1)),'k')
 hold on
 semilogx(f1,fVib3(1:length(f1)),'Color',[0.84              0.16
0],'LineStyle',':','LineWidth',2)
  hold on
 semilogx(f1,fVib5(1:length(f1)),'b','LineStyle','--
','LineWidth',1),

 xlabel(['Frequency (Hz)',sprintf('\n'),'[a]'], 'FontSize',13),
 ylabel('Magnitude (dB)','FontSize',13),axis([1e0 1e1 -30 30]),
 set(gca,'YTick',[-40 -20 0 20],'FontSize',13),
  legend(('PS','SK','SKDT'),'Location','NorthWest','FontSize',13)
   title('Pitch angular acceleration at 0 to 10 Hz','Color',[1 0
1],'FontSize',10)

  figure(54)
 semilogx(f1,fVib2(1:length(f1)),'k')
 hold on
 semilogx(f1,fVib3(1:length(f1)),'Color',[0.84              0.16
0],'LineStyle',':','LineWidth',2)

  hold on
 semilogx(f1,fVib5(1:length(f1)),'b','LineStyle','--
','LineWidth',1),

  xlabel(['Frequency (Hz)',sprintf('\n'),'[b]'], 'FontSize',13),
  ylabel('Magnitude (dB)','FontSize',13),axis([1e-1 1e2 -40 25]),
 set(gca,'YTick',[-40 -20 0 20],'FontSize',13),
  legend(('PS','SK','SKDT'),'Location','NorthWest','FontSize',13)
 title('Pitch    angular    acceleration    at    wide    frequency
range','Color',[1 0 1],'FontSize',10)

% roll acceleration

fTmp    =    fft((FullVehicle.signals.values(:,3)),NFFT)/length
(FullVehicle.time);
fVib2 = 20*log10(2*abs(fftshift(fTmp)));

fTmp3  =  fft((ReSampledSkyhook4.Data),NFFT)/length(FullVehicle.
time);
fVib3 = 20*log10(2*abs(fftshift(fTmp3)));
```

```
fTmp5     =     fft((FullVehicle.signals.values(:,4)),NFFT)/length
(FullVehicle.time);
fVib5 = 20*log10(2*abs(fftshift(fTmp5)));

figure(55)
 semilogx(f1,fVib2(1:length(f1)),'k')
 hold on
 semilogx(f1,fVib3(1:length(f1)),'Color',[0.84                0.16
0],'LineStyle',':','LineWidth',2)
  hold on
             semilogx(f1,fVib5(1:length(f1)),'b','LineStyle','--
','LineWidth',1),

   xlabel(['Frequency (Hz)',sprintf('\n'),'[a]'], 'FontSize',13),
  ylabel('Magnitude (dB)','FontSize',13),axis([1e0 1e1 -30 30]),
 set(gca,'YTick',[-40 -20 0 20],'FontSize',13),
  legend(('PS','SK','SKDT'),'Location','NorthWest','FontSize',13)
   title('Roll angular acceleration at 0 to 10 Hz','Color',[1 0
1],'FontSize',10)

  figure(56)
 semilogx(f1,fVib2(1:length(f1)),'k')
 hold on
 semilogx(f1,fVib3(1:length(f1)),'Color',[0.84                0.16
0],'LineStyle',':','LineWidth',2)
  hold on
             semilogx(f1,fVib5(1:length(f1)),'b','LineStyle','--
','LineWidth',1),

  xlabel(['Frequency (Hz)',sprintf('\n'),'[b]'], 'FontSize',13),
  ylabel('Magnitude (dB)','FontSize',13),axis([1e-1 1e2 -45 45]),
 set(gca,'YTick',[-40 -20 0 20 40],'FontSize',13),
  legend(('PS','SK','SKDT'),'Location','NorthWest','FontSize',13)
     title('Roll   angular   acceleration   at   wide   frequency
range','Color',[1 0 1],'FontSize',10)
```

Appendix C

MATLAB Code for Time Domain Analysis of Full Car Model

```
clc
clear all
close all

load('C:\Users\s\Dropbox\Full      car      cornering      160512
\roadprofileClassABC20sEach280512.mat')

SpecifiedStopTime = 10;

 RoadSteerBank280512;
 gcs;
 set_param('RoadSteerBank280512', 'StopTime', 'SpecifiedStopTime');
 sim('C:\Users\s\Dropbox\Full      car      cornering      160512
\RoadSteerBank280512.mdl')

% set_param('RoadSteerBank280512', 'SimulationCommand', 'start');
% pause(SpecifiedStopTime);
% roadprofile = PlainRoad;
% %  roadprofile = roadprofilePaper;
%   roadprofile = roadprofileClassA;
% roadprofile = roadprofileClassB;
 roadprofile = roadprofileClassC;

% roadprofile = roadprofileClassABC20sEach280512; %change simula-
tion time of the full model = dsaiDes;
```


S. Kashem et al., *Vehicle Suspension Systems and Electromagnetic Dampers*,
Springer Tracts in Mechanical Engineering, DOI 10.1007/978-981-10-5478-5

```
% Delta = SteerZero;
% Beta = BetaZero;
% %
% Delta = Steer1;
% Beta = BetaZero;
% %
% % Delta = SteerZero;
% % Beta = Beta1;
% %
% Delta = Steer1;
% Beta = Beta1;
% % %
Delta = Steer2;
Beta = Beta2;

%% ****************************************
%% First of all, chane the stop time to 10 then
%% **********************************************
full_carstabilizedcModelz1z2z3z4usedSfunc160512;
gcs;
set_param('full_carstabilizedcModelz1z2z3z4usedSfunc160512',
'StopTime', 'SpecifiedStopTime');
%    set_param('full_carstabilizedcModelz1z2z3z4usedSfunc160512',
'SimulationCommand', 'start');

full_carstabilizedcModelz1z2z3z4usedSfunc160512skyhook;
gcs;
set_param('full_carstabilizedcModelz1z2z3z4usedSfunc160512skyhook',
'StopTime', 'SpecifiedStopTime');
% set_param('full_carstabilizedcModelz1z2z3z4usedSfunc160512skyhook',
'SimulationCommand', 'start');

full_carstabilizedcModelz1z2z3z4usedSfunc160512DTC;
gcs;
set_param('full_carstabilizedcModelz1z2z3z4usedSfunc160512DTC',
'StopTime', 'SpecifiedStopTime');
%  set_param('full_carstabilizedcModelz1z2z3z4usedSfunc160512DTC',
'SimulationCommand', 'start');
% pause(SpecifiedStopTime);

sim('C:\Users\s\Dropbox\Full      car      cornering      160512
\full_carstabilizedcModelz1z2z3z4usedSfunc160512.mdl')
sim('C:\Users\s\Dropbox\Full      car      cornering      160512
\full_carstabilizedcModelz1z2z3z4usedSfunc160512skyhook.mdl')
```

```
sim('C:\Users\s\Dropbox\Full        car        cornering        160512
\full_carstabilizedcModelz1z2z3z4usedSfunc160512DTC.mdl')

%% Now resample the outputs and compare
SkyhookTimeSeriesOutput  =  timeseries(FullVehicleSkyhook.signals.
values(:,2),FullVehicleSkyhook.time);
ReSampledSkyhook2 = resample(SkyhookTimeSeriesOutput,FullVehicle.
time);

SkyhookTimeSeriesOutput  =  timeseries(FullVehicleSkyhook.signals.
values(:,4),FullVehicleSkyhook.time);
ReSampledSkyhook4 = resample(SkyhookTimeSeriesOutput,FullVehicle.
time);

SkyhookTimeSeriesOutput  =  timeseries(FullVehicleSkyhook.signals.
values(:,6),FullVehicleSkyhook.time);
ReSampledSkyhook6 = resample(SkyhookTimeSeriesOutput,FullVehicle.
time);

SkyhookTimeSeriesOutput  =  timeseries(FullVehicleSkyhook.signals.
values(:,12),FullVehicleSkyhook.time);
ReSampledSkyhook12       =       resample(SkyhookTimeSeriesOutput,
FullVehicle.time);

SkyhookTimeSeriesOutput  =  timeseries(FullVehicleSkyhook.signals.
values(:,13),FullVehicleSkyhook.time);
ReSampledSkyhook13       =       resample(SkyhookTimeSeriesOutput,
FullVehicle.time);

SkyhookTimeSeriesOutput  =  timeseries(FullVehicleSkyhook.signals.
values(:,14),FullVehicleSkyhook.time);
ReSampledSkyhook14       =       resample(SkyhookTimeSeriesOutput,
FullVehicle.time);

SkyhookTimeSeriesOutput  =  timeseries(FullVehicleSkyhook.signals.
values(:,15),FullVehicleSkyhook.time);
ReSampledSkyhook15       =       resample(SkyhookTimeSeriesOutput,
FullVehicle.time);

SkyhookTimeSeriesOutput  =  timeseries(FullVehicleSkyhook.signals.
values(:,16),FullVehicleSkyhook.time);
ReSampledSkyhook16       =       resample(SkyhookTimeSeriesOutput,
FullVehicle.time);
```

```
SkyhookTimeSeriesOutput = timeseries(FullVehicleSkyhook.signals.
values(:,18),FullVehicleSkyhook.time);
ReSampledSkyhook18       =      resample(SkyhookTimeSeriesOutput,
FullVehicle.time);

SkyhookTimeSeriesOutput = timeseries(FullVehicleSkyhook.signals.
values(:,20),FullVehicleSkyhook.time);
ReSampledSkyhook20       =      resample(SkyhookTimeSeriesOutput,
FullVehicle.time);

SkyhookTimeSeriesOutput = timeseries(FullVehicleSkyhook.signals.
values(:,21),FullVehicleSkyhook.time);
ReSampledSkyhook21       =      resample(SkyhookTimeSeriesOutput,
FullVehicle.time);

SkyhookTimeSeriesOutput = timeseries(FullVehicleSkyhook.signals.
values(:,26),FullVehicleSkyhook.time);
ReSampledSkyhook26       =      resample(SkyhookTimeSeriesOutput,
FullVehicle.time);

%% compare 'Delta','Beta','faiD'},'FontSize',14);
  figure (8)
plot(FullVehicle.time,(          FullVehicle.signals.values(:,15)
),'Color',[1 0 1],'LineStyle','--','LineWidth',2)
hold on
plot(FullVehicle.time,(          FullVehicle.signals.values(:,28)
),'k','LineStyle',':','LineWidth',2)
hold on
plot(FullVehicle.time,( FullVehicle.signals.values(:,16) ),'b')

xlabel(['Time     (s)',sprintf('\n'),'[a]'],'FontSize',14),ylabel
('Angle (radian) ','FontSize',14), %chnage
axis([0 10 -.3 .3]),
set(gca,'XTick',[0 2 4 6 8 10],'FontSize',14)
legend({'Delta','Beta','faiD'},'FontSize',14);%chnage
%% compare {'f1tilt','f4tilt'}DTC
       figure (7)
 plot(FullVehicle.time,(         FullVehicle.signals.values(:,13)
),'k','LineStyle','--','LineWidth',1)
hold on
plot(FullVehicle.time,( FullVehicle.signals.values(:,14) ),'b')

xlabel(['Time     (s)',sprintf('\n'),'[b]'],'FontSize',14),ylabel
('Force (N)','FontSize',14), %chnage
```

```
axis([0 10 -60 60]),
set(gca,'XTick',[0 2 4 6 8 10],'FontSize',14)
legend({'f1tilt','f4tilt'},'Location','Best','FontSize',8);%
chnage
% %%saveas(gcf,'f7TimeRes110612','bmp');

%% Comparison of ddz
figure (1)
plot(FullVehicle.time,(          FullVehicle.signals.values(:,1)
),'k','LineStyle','--','LineWidth',1)
hold on
plot          (ReSampledSkyhook2.time,(    ReSampledSkyhook2.Data
),'Color',[0.84 0.16 0],'LineStyle',':','LineWidth',2)
hold on
plot(FullVehicle.time,( FullVehicle.signals.values(:,2)),'b')

xlabel(['Time     (s)',sprintf('\n'),'[a]'],'FontSize',14),ylabel
('Body Acceleration (m/s2)','FontSize',14),
legend(('PS','SK','SKDT'),'Location','Best','FontSize',8)

axis([0 10 -1.5 1.5]),
set(gca,'XTick',[0 2 4 6 8 10],'FontSize',14);

figure (111)
plot(FullVehicle.time,(          FullVehicle.signals.values(:,1)
),'k','LineStyle','--','LineWidth',1)
hold on
plot          (ReSampledSkyhook2.time,(    ReSampledSkyhook2.Data
),'Color',[0.84 0.16 0],'LineStyle',':','LineWidth',2)
hold on
plot(FullVehicle.time,( FullVehicle.signals.values(:,2) ),'b')
xlabel(['Time     (s)',sprintf('\n'),'[b]'],'FontSize',14),ylabel
('Body Acceleration (m/s2)','FontSize',14),
legend(('PS','SK','SKDT'),'Location','Best','FontSize',8)

axis([6 6.5 -1 1]),
set(gca,'XTick',[0 2 4  5 6 6.5 7 8 10],'FontSize',14);
%   %%%saveas(gcf,'f1TimeRes110612','bmp');

% RMS OF Sprung Mass  Acceleration (m/s2)./9.8
Compare1  = zeros(1,3);
Compare1(1) = sqrt(   mean   ( FullVehicle.signals.values(:,1) ./
9.8 ) .^2); %Passive Model Output
Compare1(2) = sqrt(   mean   ( ReSampledSkyhook2.Data ./ 9.8 ) .
```

```
^2); %Modified skyhook control Model output
Compare1(3) = sqrt(    mean   ( FullVehicle.signals.values(:,2)./
9.8 ) .^2); %SKDT Model output

    figure(11)
 bar(Compare1)
    set(gca,'XTickLabel',('PS','SK','SKDT'),'FontSize',14)
                ylabel('Body    Normalized    RMS    Acceleration
(m/s2)','FontSize',14)
%     %%%saveas(gcf,'f11TimeRes110612','bmp');

    Compare01  = zeros(1,3);
    Compare01(1) = ( Compare1(1)- Compare1(2) )/Compare1(1)*100;
    Compare01(2) = ( Compare1(1)- Compare1(3) )/Compare1(1)*100;
%     Compare01(3) = ( Compare1(1)- Compare1(4) )/Compare1(1)*100;
   Compare01

%% compare ddseta or pitch
figure (2)
plot(FullVehicle.time,(            FullVehicle.signals.values(:,5)
),'k','LineStyle','--','LineWidth',1)
hold on
plot           (ReSampledSkyhook6.time,(    ReSampledSkyhook6.Data
),'Color',[0.84 0.16 0],'LineStyle',':','LineWidth',2)
hold on
plot(FullVehicle.time,( FullVehicle.signals.values(:,6)),'b')

xlabel(['Time     (s)',sprintf('\n'),'[a]'],'FontSize',14),ylabel
('Pitch Angular Acceleration (rad/s2)','FontSize',14),
 legend(('PS','SK','SKDT'),'Location','Best','FontSize',8)

axis([0 10 -3 3]),
set(gca,'XTick',[0 2 4 6 8 10],'FontSize',14);

figure (222)
plot(FullVehicle.time,(            FullVehicle.signals.values(:,5)
),'k','LineStyle','--','LineWidth',1)
hold on
plot           (ReSampledSkyhook6.time,(    ReSampledSkyhook6.Data
),'Color',[0.84 0.16 0],'LineStyle',':','LineWidth',2)
hold on
plot(FullVehicle.time,( FullVehicle.signals.values(:,6) ),'b')
```

```
xlabel(['Time      (s)',sprintf('\n'),'[b]'],'FontSize',14),ylabel
('Pitch Angular Acceleration (rad/s2)','FontSize',14),
 legend(('PS','SK','SKDT'),'Location','Best','FontSize',8)

axis([6 6.5 -2 2]),
set(gca,'XTick',[0 2 4  5 6 6.5 7 8 10],'FontSize',14);
%   %%%saveas(gcf,'f2TimeRes110612','bmp')

% RMS OF Acceleration (m/s2)./9.8
    Compare2  = zeros(1,3);

Compare2(1) = sqrt(    mean   ( FullVehicle.signals.values(:,5) ./
9.8 ) .^2); %Passive Model Output
Compare2(2) =             sqrt(    mean    ( ReSampledSkyhook6.Data ./
9.8 ) .^2); %Modified skyhook control Model output
Compare2(3) = sqrt(    mean   ( FullVehicle.signals.values(:,6) ./
9.8 ) .^2); %SKDT Model output
    figure(22)
bar(Compare2)
    set(gca,'XTickLabel',('PS','SK','SKDT'),'FontSize',14)
                 ylabel('Pitch     Angular     RMS     Acceleration
(rad/s2)','FontSize',14)
%     %%saveas(gcf,'f22TimeRes110612','bmp');
    Compare02  = zeros(1,3);
    Compare02(1) = ( Compare2(1)- Compare2(2) )/Compare2(1)*100;
    Compare02(2) = ( Compare2(1)- Compare2(3) )/Compare2(1)*100;
Compare02

%% compare ddfai or roll
    figure (3)
plot(FullVehicle.time,(             FullVehicle.signals.values(:,3)
),'k','LineStyle','--','LineWidth',1)
hold on
plot          (ReSampledSkyhook4.time,(    ReSampledSkyhook4.Data
),'Color',[0.84 0.16 0],'LineStyle',':','LineWidth',2)
hold on
plot(FullVehicle.time,( FullVehicle.signals.values(:,4) ),'b')

xlabel(['Time      (s)',sprintf('\n'),'[a]'],'FontSize',14),ylabel
('Roll Angular Acceleration (rad/s2)','FontSize',14),
legend(('PS','SK','SKDT'),'Location','Best','FontSize',8)

axis([0 10 -5 5]),
set(gca,'XTick',[0 2 4 6 8 10],'FontSize',14);
```

```
 figure (333)
plot(FullVehicle.time,(           FullVehicle.signals.values(:,3)
),'k','LineStyle','--','LineWidth',1)
hold on
plot          (ReSampledSkyhook4.time,(    ReSampledSkyhook4.Data
),'Color',[0.84 0.16 0],'LineStyle',':','LineWidth',2)
hold on
plot(FullVehicle.time,( FullVehicle.signals.values(:,4) ),'b')

xlabel(['Time     (s)',sprintf('\n'),'[b]'],'FontSize',14),ylabel
('Roll Angular Acceleration (rad/s2)','FontSize',14),
legend(('PS','SK','SKDT'),'Location','SouthWest','FontSize',8)

axis([6 6.5 -4.5 3]),
set(gca,'XTick',[0 2 4  5 6 6.5 7 8 10],'FontSize',14);
% %%saveas(gcf,'f3TimeRes110612','bmp');

% RMS OF   Acceleration (m/s2)./9.8
    Compare3  = zeros(1,3);

    Compare3(1) = sqrt(   mean  ( FullVehicle.signals.values(:,3)
./ 9.8 ) .^2); %Passive Model Output
   Compare3(2) = sqrt        (   mean  ( ReSampledSkyhook4.Data ./
9.8 ) .^2); %Modified skyhook control Model output
    Compare3(3) = sqrt(   mean  ( FullVehicle.signals.values(:,4)
./ 9.8 ) .^2); %SKDT Model output
    figure(33)
bar(Compare3)
    set(gca,'XTickLabel',('PS','SK','SKDT'),'FontSize',14)
   ylabel('Roll Angular RMS Acceleration (rad/s2)','FontSize',14)
    %%saveas(gcf,'f33TimeRes110612','bmp');
    Compare03  = zeros(1,3);
    Compare03(1) = ( Compare3(1)- Compare3(2) )/Compare3(1)*100;
    Compare03(2) = ( Compare3(1)- Compare3(3) )/Compare3(1)*100;
 Compare03

%% 'Comparison of ay lateral acceleration
    figure (9)
plot(FullVehicle.time,(          FullVehicle.signals.values(:,17)
),'k','LineStyle','--','LineWidth',1)
hold on
plot         (ReSampledSkyhook18.time,(    ReSampledSkyhook18.Data
),'Color',[0.84 0.16 0],'LineStyle',':','LineWidth',2)
hold on
plot(FullVehicle.time,( FullVehicle.signals.values(:,18) ),'b')
```

```
xlabel(['Time      (s)',sprintf('\n'),'[a]'],'FontSize',14),ylabel
('Lateral Acceleration (m/s2)','FontSize',14), %chnage
axis([0 10 -5 5]),
set(gca,'XTick',[0 2 4 6 8 10],'FontSize',14)
legend(('PS','SK','SKDT'),'Location','SouthWest','FontSize',8);%
chnage

    figure (999)
plot(FullVehicle.time,(          FullVehicle.signals.values(:,17)
),'k','LineStyle','--','LineWidth',1)
hold on
plot        (ReSampledSkyhook18.time,(    ReSampledSkyhook18.Data
),'Color',[0.84 0.16 0],'LineStyle',':','LineWidth',2)
hold on
plot(FullVehicle.time,( FullVehicle.signals.values(:,18) ),'b')

xlabel(['Time      (s)',sprintf('\n'),'[b]'],'FontSize',14),ylabel
('Lateral Acceleration (m/s2)','FontSize',14), %chnage
axis([6 6.5 -5 1]),
set(gca,'XTick',[0 2 4  5 6 6.5 7 8 10],'FontSize',14)
legend(('PS','SK','SKDT'),'Location','SouthWest','FontSize',8);%
chnage

% RMS Lateral  Acceleration (m/s2)./9.8
    Compare9  = zeros(1,3);

    Compare9(1) = sqrt(   mean  ( FullVehicle.signals.values(:,17)
./ 9.8 ) .^2); %Passive Model Output
    Compare9(2) =          sqrt(   mean   ( ReSampledSkyhook18.Data
./ 9.8 ) .^2); %Modified skyhook control Model output
    Compare9(3) = sqrt(   mean  ( FullVehicle.signals.values(:,18)
./ 9.8 ) .^2); %SKDT Model output
    figure(98)
bar(Compare9)
   set(gca,'XTickLabel',('PS','SK','SKDT'),'FontSize',14)
    ylabel('Lateral RMS Acceleration (m/s2)','FontSize',14)
    %%saveas(gcf,'f98TimeRes110612','bmp');
        Compare09  = zeros(1,3);
    Compare09(1) = ( Compare9(1)- Compare9(2) )/Compare9(1)*100;
    Compare09(2) = ( Compare9(1)- Compare9(3) )/Compare9(1)*100;
Compare09

%%  Road handling performance rms/fstatic
% %Fstatic = (m1 +m)* g;
% %Fz1 = k .* (z1-q)== Tyre Loads
```

```
Compare99  = zeros(1,3);
     Compare99(1) = sqrt(   mean  ( ((FullVehicle.signals.values
(:,19))./FullVehicle.signals.values(:,21)).^2) ); %Passive Model
Output
    Compare99(2) =          sqrt(  mean  ( (ReSampledSkyhook20.
Data./ReSampledSkyhook21.Data).^2) ); %Modified skyhook control
Model output

     Compare99(3) = sqrt(   mean  ( (FullVehicle.signals.values
(:,20)./FullVehicle.signals.values(:,21)).^2) );

    figure(99)
    bar(Compare99)
    set(gca,'XTickLabel',('PS','SK','SKDT'),'FontSize',14)
   %title('Comparison of different control ')
         ylabel('RMS  Normalized  road  handling  performance
(N)','FontSize',14)
    %%saveas(gcf,'f99TimeRes110612','bmp');
     Compare099  = zeros(1,3);
     Compare099(1)  = ( Compare99(2)- Compare99(1) )/Compare99(1)
*100;
     Compare099(2)  = ( Compare99(3)- Compare99(1) )/Compare99(1)
*100;

     Compare099

%% Comparison of z1 displace
       figure (6)
              plot(FullVehicle.time,( FullVehicle.signals.values
(:,11)),'k','LineStyle','--','LineWidth',1)
hold on
plot(ReSampledSkyhook12.time,( ReSampledSkyhook12.Data ),'Color',
[0.84 0.16 0],'LineStyle',':','LineWidth',2)

hold on
              plot(FullVehicle.time,( FullVehicle.signals.values
(:,12) ),'b')

xlabel(['Time    (s)',sprintf('\n'),'[a]'],'FontSize',14),ylabel
('z1 displacement (m)','FontSize',14), %chnage
legend(('PS','SK','SKDT'),'Location','Best','FontSize',8)

axis([0 10 -.08 .08]),
set(gca,'XTick',[0 2 4 6 8 10],'FontSize',14);
%%saveas(gcf,'f6TimeRes110612','bmp');
```

```
  figure (666)
               plot(FullVehicle.time,( FullVehicle.signals.values
(:,11)),'k','LineStyle','--','LineWidth',1)
hold on
plot(ReSampledSkyhook12.time,( ReSampledSkyhook12.Data ),'Color',
[0.84 0.16 0],'LineStyle',':','LineWidth',2)
hold on
               plot(FullVehicle.time,( FullVehicle.signals.values
(:,12) ),'b')

xlabel(['Time    (s)',sprintf('\n'),'[b]'],'FontSize',14),ylabel
('z1 displacement (m)','FontSize',14), %chnage
legend(('PS','SK','SKDT'),'Location','SouthWest','FontSize',8)

axis([6 7 -.04 .02]),
set(gca,'XTick',[0 2 4 5 6 7 8 10],'FontSize',14);
%%saveas(gcf,'f6TimeRes110612','bmp');

%% Comparison of ay/g force
figure (102)
               plot(FullVehicle.time,( FullVehicle.signals.values
(:,25) ),'k','LineStyle','--','LineWidth',1)
hold on
plot(ReSampledSkyhook26.time,(  ReSampledSkyhook26.Data),'Color',
[0.84 0.16 0],'LineStyle',':','LineWidth',2)
hold on
               plot(FullVehicle.time,( FullVehicle.signals.values
(:,26)),'b')

xlabel(['Time    (s)',sprintf('\n'),'[a]'],'FontSize',14),ylabel
('Rollover threshold','FontSize',14), %chnage
legend(('PS','SK','SKDT'),'Location','SouthWest','FontSize',8)

axis([0 10 -0.7 0.7]),
set(gca,'XTick',[0 2 4 6 8 10],'FontSize',14);

figure (10222)
plot(FullVehicle.time,(          FullVehicle.signals.values(:,25)
),'k','LineStyle','--','LineWidth',1)
hold on
plot    (ReSampledSkyhook26.time,     (    ReSampledSkyhook26.Data
),'Color',[0.84 0.16 0],'LineStyle',':','LineWidth',2)
hold on
plot(FullVehicle.time,( FullVehicle.signals.values(:,26) ),'b')
```

```
xlabel(['Time     (s)',sprintf('\n'),'[b]'],'FontSize',14),ylabel
('Rollover threshold','FontSize',14), %chnage
legend(('PS','SK','SKDT'),'Location','SouthWest','FontSize',8)

axis([5 5.5 -0.2 0.5]),
set(gca,'XTick',[0 2 4  5 5.5 6 6.5 7 8 10],'FontSize',14)

%tBi2h
tBi2h = mean(FullVehicle.signals.values(:,27))

% pause(1)
% close all
```

References

1. Chen Y (2009) Skyhook surface sliding mode control on semi-active vehicle suspension systems for ride comfort enhancement. Engineering 1:23–32
2. Gao W, Zhang N, Dai J (2008) A stochastic quarter-car model for dynamic analysis of vehicles with uncertain parameters. Veh Syst Dyn Int J Veh Mech Mobil 46:1159–1169
3. Choi SB, Lee HS, Park YP (2002) H-infinity control performance of a full-vehicle suspension featuring magnetorheological dampers. Veh Syst Dyn 38(5):341–360
4. Ahmadian M, Simon DE (2003) An analytical and experimental evaluation of magneto rheological suspensions for heavy trucks. Veh Syst Dyn 37:38–49
5. Ahmadian M, Vahdati N (2006) Transient dynamics of semiactive suspensions with hybrid control. J Intell Mater Syst Struct 17:145–153
6. Yagiz N, Sakman LE (2005) Robust sliding mode control of a full vehicle without suspension gap loss. J Vib Control 11(11):1357–1374
7. Yu M, Liao C, Chen W, Huang S (2006) Study on MR semi-active suspension system and its road testing. J Intell Mater Syst Struct 17:801–806
8. Eslaminasab N, Biglarbegian M, Melek WW, Golnaraghi MF (2007) A neural network based fuzzy control approach to improve ride comfort and road handling of heavy vehicles using semi-active dampers. Int J Heavy Veh Syst 14:135–157
9. Choi S, Lee S, Park Y (2001) A hysteresis model for the field-dependent damping force of a magnetorheological damper. J Sound Vib 245:375
10. Sung KG, Han YM, Lim KH, Choi SB (2007) Discrete-time fuzzy sliding mode control for a vehicle suspension system featuring an electrorheological fluid damper. Smart Mater Struct 16:798
11. Fang X, Chen W (1999) Fuzzy control technology and the application to vehicle semi-active suspension. Chin J Mech Eng 35:98–100
12. Dong X, Yu M, Liao C, Chen W, Zhang H, Huang S (2006) Adaptive fuzzy neural network control for transient dynamics of magneto-rheological suspension with time-delay. Adv Neural Netw ISNN 1046–1051
13. Yang J, Li J, Du Y (2006) Adaptive fuzzy control of lateral semi-active suspension for high-speed railway vehicle. Comput Intell 1104–1115
14. Karnopp D, Crosby M, Harwood R (1974) Vibration control using semi-active force generators. J Eng Ind 96:619–626
15. Besinger F, Cebon D, Cole D (1995) Force control of a semi-active damper. Veh Syst Dyn 24:695–723
16. Nguyen LH, Park S, Turnip A, Hong KS (2009) Modified skyhook control of a suspension system with hydraulic strut mount. In: ICCAS-SICE 2009 – ICROS-SICE International joint conference 2009, proceedings, pp 1347–1352

S. Kashem et al., *Vehicle Suspension Systems and Electromagnetic Dampers*,
Springer Tracts in Mechanical Engineering, DOI 10.1007/978-981-10-5478-5

17. Marino R, Scalzi S, Cinili F (2007) Nonlinear PI front and rear steering control in four wheel steering vehicles. Veh Syst Dyn 45:1149–1168
18. Li B, Yu F (2009) Optimal model following control of four-wheel active steering vehicle, pp 881–886
19. Du F, Li J, Li L, Si D (2010) Robust control study for four-wheel active steering vehicle. In: International conference on electrical and control engineering (ICECE), pp 1830–1833
20. Goodall R (1999) Tilting trains and beyond-the future for active railway suspensions. 1. Improving passenger comfort. Comput Control Eng J 10:153–160
21. Goodall R (1999) Tilting trains and beyond. The future for active railway suspensions. 2. Improving stability and guidance. Comput Control Eng J 10:221–230
22. web13 (2012, Aug) http://www.dailymail.co.uk/sciencetech/article-1263509/Meet-electric-powered-Land-Glider-half-width-normal-car.html
23. web9 (2012, Aug) http://www.seriouswheels.com/cars/top-1997-Mercedes-Benz-F-300-Life-Jet-Concept.htm
24. web8 (2012, Aug) http://www.maxmatic.com/ttw_moto.htm. Available http://delphi.com/manufacturers/auto/other/ride/magneride/
25. web11 (2012, Aug) http://www.carver-engineering.com/
26. Karnopp D, Hibbard R (1992) Optimum roll angle behavior for tilting ground vehicles. ASME Dyn Syst Control Div Publ DSC, ASME, New York, NY (USA) 44:29–37
27. Hibbard R, Karnopp D (1996) Twenty first century transportation system solutions—a new type of small, relatively tall and narrow active tilting commuter vehicle. Veh Syst Dyn 25:321–347
28. So SG, Karnopp D (1993) Methods of controlling the lean angle of tilting vehicles. DSC 52:311–319
29. Saccon A, Hauser J, Beghi A (2008) A virtual rider for motorcycles: an approach based on optimal control and maneuver regulation, pp 243–248
30. Frezza R, Beghi A (2003) Simulating a motorcycle driver. In: New trends in nonlinear dynamics and control and their applications, pp 175–186
31. Kidane S, Alexander L, Rajamani R, Starr P, Donath M (2006) Road bank angle considerations in modeling and tilt stability controller design for narrow commuter vehicles. In: Proceedings of the American control conference, Minneapolis, Minnesota, USA, p 6
32. Piyabongkarn D, Keviczky T, Rajamani R (2004) Active direct tilt control for stability enhancement of a narrow commuter vehicle. Int J Automot Technol 5:77–88
33. Wang J, Shen S (2008) Integrated vehicle ride and roll control via active suspensions. Veh Syst Dyn 46:495–508
34. Jones W (2005) Easy ride: Bose Corp. uses speaker technology to give cars adaptive suspension. IEEE Spectr 42:12–14
35. web28 (2012, Aug) http://www.motortrend.com/auto_shows/tokyo/2009/
36. Gohl JB (2003) Narrow tilting vehicles: modeling and steering based, tilt control. University of Minnesota, Minnesota
37. Rajamani R, Gohl J, Alexander L, Starr P (2003) Dynamics of narrow tilting vehicles. Math Comput Model Dyn Syst 9:209–231
38. So SG, Karnopp D (1997) Switching strategies for narrow ground vehicles with dual mode automatic tilt control. Int J Veh Des 18:518–532
39. Sang-Gyun S, Karnopp D (1997) Switching strategies for narrow ground vehicles with dual mode automatic tilt control. Int J Veh Des 18:518–532
40. Li Y, Meiry J, Roeseler W (1968) An active roll mode suspension system for ground vehicles. J Basic Eng 90:167
41. Mourad L, Claveau F, Chevrel P (2011) A lateral control strategy for narrow tilting commuter vehicle based on the perceived lateral acceleration. In: 18th IFAC World Congress Milano (Italy), September 2011, pp 6254–6259

42. Roqueiro N, de Faria MG, Colet EF (2011) Sliding mode controller and flatness based set-point generator for a three wheeled narrow vehicle. In: 18th IFAC World Congress Milano (Italy), September, 2011, pp 11925–11930
43. Amati N, Festini A, Pelizza L, Tonoli A (2012) Dynamic modelling and experimental validation of three wheeled tilting vehicles. Veh Syst Dyn 49:889–914
44. Edelmann J, Plöchl M, Lugner P (2011) Modelling and analysis of the dynamics of a tilting three-wheeled vehicle. Multibody Syst Dyn 1–19
45. Jiang X, Wang J, Hu H (2012) Semi-active control of a vehicle suspension using magneto-rheological damper. J Cent South Univ 19:1839–1845
46. Chen SZ, Feng ZZ, Yang L, Zhao YT (2012) Magnetorheological semi-active suspension demonstration for off-road vehicles. Adv Sci Lett 12:1–6
47. Sun P, Wang JF, Feng F (2012) A new kind of semi-active suspension magnetorheological fluid shock absorber damping force model. Adv Mater Res 549:856–860
48. Pellegrini E, Koch G, Spirk S, Lohmann B (2011) A dynamic feedforward control approach for a semi-active damper based on a new hysteresis model, pp 6248–6253
49. Wang WR, Sun ZH, Yan XQ, Jiang Q (2012) Analysis of vane hydraulic damp static seal mechanism and performance influencing factors. Appl Mech Mater 152:1774–1778
50. ISO (1995) Mechanical vibration-road surface profiles-reporting of measured data, TC 108/SC 2 (Mechanical vibration and shock/measurement and evaluation of mechanical vibration and shocks as applied to machines, vehicles and structures), ISO 8608: E Standards, ISO
51. Nguyen QH, Choi SB (2009) Optimal design of MR shock absorber and application to vehicle suspension. Smart Mater Struct 18:035012
52. Karnopp D, Crosby MJ, Harwood RA (1973) Vibration control using semi-active force generators. In: ASME Pap
53. Hrovat D (1997) Survey of advanced suspension developments and related optimal control applications. Automatica 33:1781–1817
54. Poussot-Vassal C, Sename O, Dugard L, Gaspar P, Szabo Z, Bokor J (2008) A new semi-active suspension control strategy through LPV technique. Control Eng Pract 16:1519–1534
55. Rossi C, Lucente G (2004) H_∞ control of automotive semi-active suspensions. In: Proceedings of the 1st IFAC symposium on advances in automotive control Salerno, Italy
56. Zin A, Sename O, Dugard L (2005) Switched H_∞ control strategy of automotive active suspensions. In: Proceedings of the 16th IFAC world congress (WC), Praha, Czech Republic
57. Giorgetti N, Bemporad A, Tseng HE, Hrovat D (2006) Hybrid model predictive control application towards optimal semi-active suspension. Int J Control 79:521–533
58. Poussot-Vassal C, Drivet A, Sename O, Dugard L, Ramirez-Mendoza R (2007) A self tuning LPV/H_∞ suspension controller for a multi-body quarter vehicle model. In: Proceedings of the 10th mini-conference on vehicle system dynamics, identification and anomalies (VSDIA), Budapest, Hungary
59. Gaspar P, Szaszi I, Bokor J (2004) Active suspension design using LPV control. In: Proceedings of the 1st IFAC symposium on advances in automotive control (AAC), Salerno, Italy, pp 584–589
60. Zin A, Sename O, Gaspar P, Dugard L, Bokor J (2008) Robust LPV–H_∞ control for active suspensions with performance adaptation in view of global chassis control. Veh Syst Dyn 46:889–912
61. Canale M, Milanese M, Novara C (2006) Semi-active suspension control using 'fast' model-predictive techniques. IEEE Trans Control Syst Technol 14:1034–1046
62. Giua A, Melas M, Seatzu C, Usai G (2004) Design of a predictive semiactive suspension system. Veh Syst Dyn 41:277–300
63. Choudhury SF, Sarkar DMAR (2012) An approach on performance comparison between automotive passive suspension and active suspension system (pid controller) using matlab/simulink. J Theor Appl Inf Technol 43:295–300
64. ElMadany MM, Abduljabbar ZS (1999) Linear quadratic Gaussian control of a quarter-car suspension. Veh Syst Dyn 32:479–497

65. Krtolica R, Hrovat D (1992) Optimal active suspension control based on a half-car model: an analytical solution. IEEE Trans Autom Control 37:528–532
66. Hrovat D (1991) Optimal active suspensions for 3D vehicle models. In: Proceedings of American control conference, pp 1534–1541
67. Doyle J, Stein G (1981) Multivariable feedback design: concepts for a classical/modern synthesis. IEEE Trans Autom Control 26:4–16
68. Fuller CR, Elliott SJ, Nelson PA (1997) Active control of vibration. Academic Press, London
69. Hrovat D (1993) Applications of optimal control to advanced automotive suspension design. J Dyn Syst Meas Control 115:328
70. Shisheie R, Shafieenejad I, Moallemi N, Novinzadeh A (2012) Linear quadratic regulator time-delay controller for hydraulic actuator. J Basic Appl Sci Res 2(3):2607–2618
71. Prokop G, Sharp R (1995) Performance enhancement of limited-bandwidth active automotive suspensions by road preview. IEEE Proc Control Theory Appl 140–148
72. Shen Y (2005) Vehicle suspension vibration control with magnetorheological dampers. Ph.D. Thesis, University of Waterloo
73. Foo G, Rahman M (2010) Sensorless sliding-mode MTPA control of an IPM synchronous motor drive using a sliding-mode observer and HF signal injection. IEEE Trans Ind Electron 57:1270–1278
74. Gupta R, Ghosh A, Joshi A (2010) Multiband hysteresis modulation and switching characterization for sliding-mode-controlled cascaded multilevel inverter. IEEE Trans Ind Electron 57:2344–2353
75. Shi P, Xia Y, Liu G, Rees D (2006) On designing of sliding-mode control for stochastic jump systems. IEEE Trans Autom Control 51:97–103
76. Shiri A (2012) Robust sliding mode control of electromagnetic suspension system with parameter uncertainty. J Appl Sci Eng Technol 10:1677–1683
77. Orowska-Kowalska T, Kaminski M, Szabat K (2010) Implementation of a sliding-mode controller with an integral function and fuzzy gain value for the electrical drive with an elastic joint. IEEE Trans Ind Electron 57:1309–1317
78. Huang L, Mao X (2010) SMC design for robust H [infinity] control of uncertain stochastic delay systems. Automatica 46:405–412
79. Wu L, Ho DWC (2010) Sliding mode control of singular stochastic hybrid systems. Automatica 46:779–783
80. Choi HH (2007) LMI-based sliding surface design for integral sliding mode control of mismatched uncertain systems. IEEE Trans Autom Control 52:736–742
81. Tsai YW, Mai KH, Shyu KK (2006) Sliding mode control for unmatched uncertain systems with totally invariant property and exponential stability. J Chin Inst Eng 29:179–183
82. Chan ML, Tao C, Lee TT (2000) Sliding mode controller for linear systems with mismatched time-varying uncertainties. J Franklin Inst 337:105–115
83. Yagiz N, Yuksek I, Sivrioglu S (2000) Robust control of active suspensions for a full vehicle model using sliding mode control. JSME Int J Ser C Mech Syst Mach Elem Manuf 43:253–258
84. Chamseddine A, Noura H, Raharijaona T (2006) Control of linear full vehicle active suspension system using sliding mode techniques, pp 1306–1311
85. Al-Holou N, Joo DS, Shaout A (1994) The development of fuzzy logic based controller for semi-active suspension system. In: Proceedings of the 37th Midwest symposium on IEEE, vol 2, pp 1373–1376
86. Barr AJ, Ray J (1996) Control of an active suspension using fuzzy logic. In: Proceedings of the Fifth IEEE International conference on fuzzy systems, vol 1, pp 42–48
87. Yan B, Zhang X, Niu H (2012) Design and test of a novel isolator with negative resistance electromagnetic shunt damping. Smart Mater Struct 21:035003
88. Youssef KH, Yousef HA, Sebakhy OA, Wahba MA (2009) Adaptive fuzzy APSO based inverse tracking-controller with an application to DC motors. Expert Syst Appl 36:3454–3458

89. Khalid H, Rizvi S, Cheded L, Doraiswami R, Khoukhi A (2010) A PSO-trained adaptive neuro-fuzzy inference system for fault classification. In: International conference on neural computation, ICNC Valencia, Spain
90. Kashani R, Strelow JE (1999) Fuzzy logic active and semi-active control of off-road vehicle suspensions. Veh Syst Dyn 32:409–420
91. Shen Y, Golnaraghi M, Heppler G (2006) Semi-active vibration control schemes for suspension systems using magnetorheological dampers. J Vib Control 12:3–24
92. Karnopp D (1983) Active damping in road vehicle suspension systems. Veh Syst Dyn 12:291–311
93. Li H, Goodall RM (1999) Linear and non-linear skyhook damping control laws for active railway suspensions. Control Eng Pract 7:843–850
94. Margolis DL (1983) Semi-active control of wheel hop in ground vehicles. Veh Syst Dyn 12:317–330
95. Savaresi SM, Silani E, Bittanti S (2005) Acceleration-driven-damper (ADD): an optimal control algorithm for comfort-oriented semiactive suspensions. J Dyn Syst Meas Control 127:218
96. Savaresi SM, Spelta C (2007) Mixed sky-hook and ADD: approaching the filtering limits of a semi-active suspension. J Dyn Syst Meas Control 129:382
97. Bakar SAA, Jamaluddin H, Rahman RA, Samin PM, Hudha K (2008) Vehicle ride performance with semi-active suspension system using modified skyhook algorithm and current generator model. Int J Veh Auton Syst 6:197–221
98. Miller LR, Nobles CM (1990) Methods for eliminating jerk and noise in semi-active suspensions, presented at the SAE 1990 transactions
99. Tong RT (2001) Ride control: a two-state design for heavy vehicle suspension. University of Illinois at Chicago
100. Soliman AMA, Kaldas MMS, Barton DC, Brooks PC (2012) Fuzzy-skyhook control for active suspension systems applied to a full vehicle model. Int J Eng Technol 2:01–12
101. Islam ASMS, Ahmed A (2005) Comparative performance study of magneto-rheological fluid based damper for vehicle suspension. In: Proceedings of the International conference on mechanical engineering 2005, Dhaka, Bangladesh
102. Kashem SBA, Ektesabi M, Nagarajah R (2012) Comparison between different sets of suspension parameters and introduction of new modified skyhook control strategy incorporating varying road condition. Veh Syst Dyn 50:1173–1190
103. Valasek M, Kortum W, Sika Z, Magdolen L (1998) Development of semi-active road-friendly truck suspensions. Control Eng Pract 6:735–744
104. Yi K, Wargelin M, Hedrick JK, University of California Transportation Center, B. D. o. M. E (1992) Dynamic tire force control by semi-active suspensions. University of California Transportation Center, University of California
105. Valášek M, Novak M, Åika Z, Vaculin O (1997) Extended ground-hook-new concept of semi-active control of truck's suspension. Veh Syst Dyn 27:289–303
106. web7 (2012, Aug) http://www.conceptcarz.com/vehicle/z11633/Carver-One.aspx
107. web16 (2012, Aug) http://lynk.ly/stories/view/1479328. Available http://delphi.com/manufacturers/auto/other/ride/magneride/
108. web17 (2012, Aug) http://www.i4u.com/9754/carver-one-first-tilting-three-wheeler-market. Available http://delphi.com/manufacturers/auto/other/ride/magneride/
109. Rajamani R (2006) Vehicle dynamics and control. Springer, New York
110. web14 (2012, Aug) http://www.bmwblog.com/2009/10/09/bmw-unveils-clever-concept/
111. web12 (2012, Aug) http://www.MP3.piaggio.com/. Available http://delphi.com/manufacturers/auto/other/ride/magneride/
112. Snell A (1998) An active roll-moment control strategy for narrow tilting commuter vehicles. Veh Syst Dyn 29:277–307

113. Poelgeest A, Edge K, Darling J (2007) Development of a steer tilt controller for a three wheeled tilting vehicle. In: Proceedings of the ASME International mechanical engineering congress and exposition, IMECE, Seattle, USA
114. Kidane S, Alexander L, Rajamani R, Starr P, Donath M (2008) A fundamental investigation of tilt control systems for narrow commuter vehicles. Veh Syst Dyn 46:295–322
115. Chiou J, Lin CY, Chen CL, Chien CP (2009) Tilting motion control in narrow tilting vehicle using double-loop PID controller. In: Proceedings of the 7th Asian control conference, 978-89-956056-1/09, pp 913–918
116. Defoort M, Murakami T (2009) Sliding-mode control scheme for an intelligent bicycle. IEEE Trans Ind Electron 56:3357–3368
117. Nenner U, Linker R, Gutman PO (2008) Robust stabilization of an unmanned motorcycle. In: Robotics, automation and mechatronics, IEEE conference on, Chengdu, China, pp 101–106
118. web19 (2012, Aug) http://driftjapan.com/blog/car-parts/drift-car-suspension/. Available http://delphi.com/manufacturers/auto/other/ride/magneride/
119. Suda Y, Shiiba T, Hio K, Kawamoto Y, Kondo T, Yamagata H (2004) Study on electromagnetic damper for automobiles with nonlinear damping force characteristics: (road test and theoretical analysis). Veh Syst Dyn 41:637–646
120. Gillespie T (2006) Development of semi-active damper for heavy off-road military vehicles. University of Waterloo
121. Irmscher S, Hees E (1966) Experience in semi-active damping with state estimators. In: Proceeding of AVEC, pp 193–206
122. Konik D, Bauer W, Huber K, Jordan B, Kolbel S, Scharf J et al (1996) Electronic damping control with continuously working damping valves (EDCC) – system description and functional requirements. In: International symposium on advanced vehicle control, Aachen, Germany, pp 87–104
123. Nakayama T, Morita M, Kamimae H, Nishihara A, Tuda K (1996) Development of semi-active control system with PUDLIS. In: Proceedings of International symposium on advanced vehicle control, AVEC'96, Aachen University of Technology, pp 233–239
124. Yi K, Song BS (1999) A new adaptive sky-hook control of vehicle semi-active suspensions. Proc Inst Mech Eng Part D J Autom Eng 213:293–303
125. Sankaranarayanan V, Emekli ME, Gilvenc B, Guvenc L, Ozturk E, Ersolmaz E et al (2008) Semiactive suspension control of a light commercial vehicle. IEEE/ASME Trans Mechatron 13:598–604
126. Zhang H, Winner H, Li W (2009) Comparison between skyhook and minimax control strategies for semi-active suspension system. World Acad Sci Eng Technol 624–627
127. Abdalla MO, Al Shabatat N, Al Qaisi M (2009) Linear matrix inequality based control of vehicle active suspension system. Veh Syst Dyn Int J Veh Mech Mobil 47:121–134
128. Fateh MM, Alavi SS (2009) Impedance control of an active suspension system. Mechatronics 19:134–140
129. Faheem A, Alam F, Thomas V (2006) The suspension dynamic analysis for a quarter car model and half car model. In: Presented at the 3rd BSME-ASME International conference on thermal engineering, Dhaka, 20–22 December 2006
130. Gupta A, Jendrzejczyk J, Mulcahy T, Hull J (2006) Design of electromagnetic shock absorbers. Int J Mech Mater Des 3:285–291
131. Guo D, Hu H, Yi J (2004) Neural network control for a semi-active vehicle suspension with a magnetorheological damper. J Vib Control 10:461–471
132. Priyandoko G, Mailah M, Jamaluddin H (2009) Vehicle active suspension system using skyhook adaptive neuro active force control. Mech Syst Signal Process 23:855–868
133. Scheibe F, Smith MC (2009) Analytical solutions for optimal ride comfort and tyre grip for passive vehicle suspensions. Veh Syst Dyn Int J Veh Mech Mobil 47:1229–1252
134. Kruczek A, Stribrsky A (2004) A full-car model for active suspension – some practical aspects. In: Proceedings of the IEEE International conference on mechatronics 2004, ICM'04, pp 41–45

135. Jia-ling Y, Jia-qiang Z (2006) Semi-active suspension system design for quarter-car model using model reference sliding mode control. In: IEEE International conference on vehicular electronics and safety, ICVES 2006, pp 398–402
136. Gopala Rao LVV, Narayanan S (2009) Sky-hook control of nonlinear quarter car model traversing rough road matching performance of LQR control. J Sound Vib 323:515–529
137. Du H, Zhang N, Lam J (2008) Parameter-dependent input-delayed control of uncertain vehicle suspensions. J Sound Vib 317:537–556
138. Sung K, Han Y, Sohn J, Choi S (2008) Road test evaluation of vibration control performance of vehicle suspension featuring electrorheological shock absorbers. Proc Inst Mech Eng Part D J Autom Eng 222:685–698
139. Collette C, Preumont A (2010) High frequency energy transfer in semi-active suspension. J Sound Vib 329:4604–4616
140. Shamsi A, Choupani N (2008) Continuous and discontinuous shock absorber control through skyhook strategy in semi-active suspension system (4DOF model). In: Proceedings of World Academy of Science, Engineering and Technology, pp 746–750
141. Taniguchi O (1976) Vibration engineers' handbook. Yokendo Press, Tokyo
142. Abramov S, Mannan S, Durieux O (2009) Semi-active suspension system simulation using Simulink. Int J Eng Syst Model Simul 1:101–114
143. ISO (1997) Mechanical vibration and shock: evaluation of human exposure to whole-body vibration. Part 1, general requirements: International Standard ISO 2631-1: (E), ISO
144. Faulhaber DC MOTOR Technical information, Faulhaber DC motor specs, Germany (2011)
145. Ikenaga S, Lewis FL, Campos J, Davis L (2000) Active suspension control of ground vehicle based on a full-vehicle model. In: Proceedings of the American control conference, pp 4019–4024
146. Ikenaga S, Lewis FL, Davis L, Campos J, Evans M, Scully S (1999) Active suspension control using a novel strut and active filtered feedback: design and implementation. In: IEEE conference on control applications – proceedings, pp 1502–1508
147. Sang-Gyun S, Karnopp D (1997) Active dual mode tilt control for narrow ground vehicles. Veh Syst Dyn 27:19–36
148. Allen RW, Rosenthal TJ, Klyde DH, Hogue JR (1999) Computer simulation analysis of light vehicle lateral/directional dynamic stability. SAE Trans 108:482–497

Lightning Source UK Ltd.
Milton Keynes UK
UKHW021333301018
331460UK00001B/31/P